KU-205-687

Agent-Based Manufacturing and Control Systems

New Agile
Manufacturing
Solutions for Achieving
Peak Performance

Agent-Based Manufacturing and Control Systems

New Agile Manufacturing Solutions for Achieving Peak Performance

Massimo Paolucci
Universita de Genova
Genova, Italy
Roberto Sacile
Universita de Genova
Genova, Italy

CRC PRESS

Boca Raton   London   New York   Washington, D.C.

Library of Congress Cataloging-in-Publication Data

Paolucci, Massimo, 1961-
   Agent-based manufacturing and control systems : new agile manufacturing solutions for achieving peak performance / Massimo Paolucci and Roberto Sacile.
      p.   cm.
   Includes bibliographical references and index.
   ISBN 1-57444-336-4 (alk. paper)
   1. Computer integrated manufacturing systems. 2. Expert systems (Computer science). 3. Manufacturing processes—Automation. 4. Production control—Data processing. I. Sacile, Roberto. II. Title.

TS155.63 2004
670′.285—dc22   2004054439

This book contains information obtained from authentic and highly regarded sources. Reprinted material is quoted with permission, and sources are indicated. A wide variety of references are listed. Reasonable efforts have been made to publish reliable data and information, but the author and the publisher cannot assume responsibility for the validity of all materials or for the consequences of their use.

Neither this book nor any part may be reproduced or transmitted in any form or by any means, electronic or mechanical, including photocopying, microfilming, and recording, or by any information storage or retrieval system, without prior permission in writing from the publisher.

The consent of CRC Press LLC does not extend to copying for general distribution, for promotion, for creating new works, or for resale. Specific permission must be obtained in writing from CRC Press LLC for such copying.

Direct all inquiries to CRC Press LLC, 2000 N.W. Corporate Blvd., Boca Raton, Florida 33431.

Trademark Notice: Product or corporate names may be trademarks or registered trademarks, and are used only for identification and explanation, without intent to infringe.

Visit the CRC Press Web site at www.crcpress.com

© 2005 by CRC Press LLC

No claim to original U.S. Government works
International Standard Book Number 1-57444-336-4
Library of Congress Card Number 2004054439
Printed in the United States of America 1 2 3 4 5 6 7 8 9 0
Printed on acid-free paper

DEDICATION

To Ciaccia and Anna

Massimo

To Cristina, Federico, and his little sister to come

Roberto

PREFACE

Rushing toward total globalization, today's market environment is characterized by an ever increasing pace in production, decreasing product cycle times, and an increasing shift from mass production to mass customization. Businesses that are more responsive to market changes and more sensitive to customer needs are more likely to survive and thrive in this kind of environment. These new circumstances require a different approach to management software systems as well. Particularly in the manufacturing sector, new management software architectures must be introduced. Decentralized, or heterarchical, management has recently been proposed as a way to overcome the limitations of traditional hierarchical or centralized information systems in such a highly dynamic environment.

In this context, the aim of this book is to introduce the reader to the world of the (autonomous) *agents* that gradually, but inexorably, will substitute for centralized information systems. Agents are software entities that have a set of protocols to govern the operations of corresponding manufacturing entities, knowledge bases, inference mechanisms, and explicit models of the problems to solve. Agents communicate and negotiate with other agents, as well as perform operations based on locally available information and possibly pursue its local goals.

The idea of agent-based manufacturing has received a great deal of attention since 1992 when the Japanese Intelligent Manufacturing Systems Program decided to make agent-based manufacturing one of its initial test case studies. This interest was galvanized even more in 1993 when the U.S. National Center for Manufacturing Science decided to initiate a number of programs in agent-based manufacturing. In fact, agent technology can be considered one of the foundation technologies for implementing the agile manufacturing vision that aspires to total flexibility without sacrificing quality or incurring added cost.

This book helps the reader to understand how an appropriate selection of entities to be modeled as agents and the definition of suitable negotiation protocols allow an agent-based system to yield extremely flexible, robust, adaptive, and fault-tolerant global performance for a given manufacturing system. This is exactly what is needed to satisfy the demands of today's market. In one sentence: *Agent-Based Manufacturing and Control Systems: New Agile Manufacturing Solutions for Achieving Peak Performance* is a comprehensive guide for researchers and practitioners to understand and design an agent-based management software for manufacturing. This book is intended to meet the needs of researchers and practitioners. In particular, its main targets are people from the academic world beginning their research into agent-based manufacturing and people from the industrial world seeking new, but established, ideas in order to organize a new manufacturing business or to reorganize an existing one to better meet market challenges.

This book surveys the literature and real-world applications. It presents techniques applicable to real-world problems by introducing a didactic, but realistic, example of a manufacturing firm. It introduces the reader to the design of agent-based systems. In particular, it indicates when an agent-based approach is useful, i.e., which problems are more suitable to be resolved using autonomous agents. Then, for a fixed problem, it proposes a possible breakdown in order to associate an agent to each basic entity of the problem. Finally, it proposes the possible properties of agents and of the negotiation protocols.

Raffaele Pesenti

ACKNOWLEDGMENTS

This work is the result of several collaborations with the industrial and academic worlds. In this respect, the authors wish to thank the following colleagues and friends: Antonio Boccalatte, Mauro Coccoli, Paolo Copello, Rose Dieng, Matteo Gentile, Andrea Gozzi, Alberto Grosso, Riccardo Minciardi, Ernesto Montaldo, Raffaele Pesenti, Agostino Poggi, Arianna Poggi, Germana Portella, Marco Repetto, Christian Vecchiola, and Riccardo Zoppoli. A special thank-you to Thomas John Wiley for his support in the revision of the manuscript.

In addition, the authors wish to thank the Department of Communication, Computer and System Sciences and Siemens–Orsi Automation SpA, without whose support the work could not have been achieved.

AUTHORS

Massimo Paolucci, Ph.D., born in Genova, Italy, graduated from the University of Genova as an electronics engineer in 1986. He received the Ph.D. in electronics and computer science at the University of Genova in 1990. Since 1992, he has been working as an assistant professor in the Department of Communication, Computer and System Sciences (DIST) of the University of Genova, teaching courses in the operations research and computer science fields.

Dr. Paolucci has worked in the fields of decision support systems; multicriteria decision-making; production scheduling in manufacturing systems; and scheduling and simulation in underground railway transportation systems. He has been involved in national and European Union projects relevant to industrial automation and environment, in particular dealing with decision-making aspects. His main research interests are currently in manufacturing and logistics systems; in particular, he is addressing job scheduling and vehicle routing problems by means of agent-based approaches. He is author of many journal and conference papers and also has professional experiences in the field of information systems and database management. He is the scientific coordinator for an academic–industrial project on agent-based scheduling funded by Siemens SpA.

Roberto Sacile, Ph.D., received a Laurea degree in electronics engineering at the University of Genova, Italy, in 1990 and a Ph.D. at Politecnico of Milano, Italy, in 1994. He was a postdoctoral fellow at INRIA (Institut National de Recherche en Informatique et en Automatique), Sophia Antipolis, France, in 1995 in the ACACIA (Knowledge Acquisition by Agent Techniques) project. He has been an invited professor in computer networks at Parma University (1994 to 1998) and in information-based models of biological systems at Genova University (1997 to 1999). Since 1999, Dr. Sacile has collaborated with ORSI group (now Siemens SpA) on agent technologies and knowledge-based systems applied to manufacturing. Since 2000, he has taught courses in computer science and system engineering as an assistant professor on the engineering faculty of the University of Genova, where he is member of the Department of Communication, Computer and System Sciences (DIST). He is currently involved in a national project on models for optimization, control, and coordination of distributed production systems sponsored by the Italian Ministry for Education, University and Research.

Dr. Sacile's research activity mainly focuses on decision support systems and agent-based techniques, with special reference to manufacturing, environmental, and healthcare information systems and applications. He is author of more than 100 papers published in peer-reviewed journals and congress proceedings.

CONTRIBUTORS

Massimo Cossentino
ICAR/CNR
(Istituto per il Calcolo Automatico e Reti ad
  Alte Prestazioni/Consiglio Nazionale delle
  Ricerche)
Palermo, Italy

Manuel Gentile
Italian National Research Council
Institute for Educational Technologies
Palermo, Italy

Raffaele Pesenti
DINFO, Dipartimento di Ingegneria Infor-
  matica University of Palermo
Palermo, Italy

Luca Sabatucci
ICAR/CNR
(Istituto per il Calcolo Automatico e Reti ad
  Alte Prestazioni/Consiglio Nazionale delle
  Ricerche)
Palermo, Italy

CONTENTS

1

AGENT TECHNOLOGY IN MODERN PRODUCTION SYSTEMS

After a preface on current trends in modern production systems, in which we define production models such as flexible, agile, and holonic manufacturing, the concepts of agent and multiagent systems (deriving from distributed artificial intelligence, as well as from decision and information technology) are introduced as a natural requisite to achieve the peak performance demanded by today's modern manufacturing. The reader is provided with the fundamental information needed to recognize the domain characteristics that indicate the appropriateness of an agent-based solution. In this respect, one of the most crucial aspects of "agility" in modern production systems seems to be the ability to manage information about production, market, and business processes effectively and promptly, i.e., to be able to manufacture it into more fruitful integrated information. In other words, information in modern production systems must be processed throughout the whole supply chain, similarly to materials and products, because it is itself a sort of material needed by the company's industrial processes.

Multiagent systems seem to provide one of the most promising technologies to render modern production systems "agile." How agility can be achieved by making agents work effectively in a modern production system is the main goal of this book.

INTRODUCTION

The manufacturing industry has entered an era in which computer technology has refocused attention from hardware platforms to operating

1

systems and software components. The need for continuous real-time information flow (available at any time to many people) is currently pushing information technology providers to develop control system models, management systems, and software designed to support this flow.

Over the last 20 years, integrated information systems have undeniably helped to solve most of the issues concerning process management and control; reducing production costs; improving process capabilities and yields; monitoring and controlling process status; and enhancing overall production. Nevertheless, the management of an effective and modern manufacturing industry should not only invest in systems that perform better and provide more profitable ways to exploit business processes, but should also encompass wider horizons, embracing all of the critical business layers within and beyond the company's boundaries.

Today's marketplace is increasingly more demanding in terms of lower costs, faster time-to-market, and better quality, thus forcing companies to become ever more reactive and agile in performing their daily business management tasks. Some manufacturing industries have founded their businesses on shorter life-cycle products or have diversified into more competitive markets in different industrial sectors. The most direct implication of this evolution is that modern manufacturing companies should be able to act like cells in an organism (the market). In simple terms, the business model is changing from open competition to one in which, for the organism to survive, strong, effectively linked cooperation among businesses — horizontally and vertically — is fundamental.

With the advent of the postindustrial age, the survival of manufacturing companies has become increasingly more dependent on their ability to react promptly and flexibly to market variations and needs. In this respect, flexibility would appear to be the strategic success factor to satisfy the global competition needs of worldwide manufacturing enterprises, allowing them to provide high-quality production at reasonable costs. Modern production systems must be distinguished by their organization of management, communication, and production tasks, as well as by planning and decisional capabilities, which allow them to rapidly respond to (or better, to predict) market needs while effectively competing on the market. "Flexibility" has become a priority objective for companies working in markets characterized by a heavily differentiated and dynamic demand.

Flexibility in manufacturing [1] is related to the ability to organize and reorganize production resources efficiently from a standpoint of price, quality, and time [2] in response to changes in the environment and, in particular, to changes in demand and technology. The need for flexibility (meant as reconfigurability and adaptability of the productive systems structures) gave rise to the concept of "holons" and holonic manufacturing systems (HMS). The term *holon*, first introduced by Koestler in 1967 [3], comes from the Greek words "holos" (meaning the whole) and "on" (the particle). Holons refer to any component of a complex system that, even when contributing to the functioning of the whole, demonstrates autonomous, stable, and self-contained behavior or function.

In manufacturing, the term *holonic* is used to stress the concept of highly decentralized coordination and control in production systems [4]. In this respect, an HMS has a structure in which functions are hierarchically distributed to autonomous entities (the holons) corresponding to specific identifiable parts of a manufacturing system, which can be made up of other subordinate parts and can, in turn, also be part of a larger whole. Experts from several countries, as well as public and private institutions, have partnered to form an HMS consortium (http//hms.ifw.uni-hannover.de/) to define the specifications for HMSs as a project in the study of intelligent manufacturing systems (IMS). According to the definition of the HMS Consortium "a holon is an autonomous and cooperative building block of a manufacturing system for transforming, transporting, storing and/or validating information and physical objects," [5] and an HMS is "a holarchy (a system of holons which can cooperate to achieve a goal or objective) which integrates the entire range of manufacturing activities from order booking through design, production and marketing to realize agile manufacturing enterprises."

It is easy to understand how concepts such as flexible manufacturing systems (FMS) [6] and computer integrated manufacturing (CIM) [7, 8], which characterized the approach to manufacturing in the 1980s and 1990s, evolved into the HMS paradigm. CIM represented the introduction of a strong vertical integration of manufacturing subsystems into a hierarchical structure allowing information needed for production to flow from the top levels of an enterprise to the shop floor. HMS moves toward a more flexible integration of functionalities among distributed autonomous actors.

Competitive factors and a greater attention to the volatility of consumer preferences, in particular as these have an impact on reducing the product life cycle, have contributed to the development of the concept of "agile production" [9]. Agile manufacturing (AM) is a more recent evolution of the previously mentioned concepts. It extends the concept of flexibility beyond manufacturing system boundaries into the environment and the marketplace, where management of customer relations and cooperation among companies are even more important.

"Agility" requires the efficient and effective utilization of internal and external resources to meet changing customer needs quickly and flexibly [10]. AM should lower manufacturing costs; increase market share; satisfy customer requirements; facilitate the rapid introduction of new products; eliminate non-value-added activities; and increase manufacturing competitiveness [11, 12].

The concepts of AM and HMS are strictly interwoven, and many definitions highlight this fact. For Christensen [13] and Deen [14], an HMS is also depicted as "a manufacturing system where key elements, such as raw materials, machines, products, parts, AGVs, etc., have autonomous and cooperative properties." For Shen and Norrie [15], the integration of the entire range of autonomous and cooperative manufacturing activities (from order booking through design, production, and marketing) allows an enterprise to achieve agile manufacturing processes.

Several articles and books have attempted to explain "how to make a company agile." According to the definition by Cho et al. [11], AM is "the capability of surviving and prospering in the competitive environment of continuous and unpredictable change by reacting quickly and effectively to changing markets, driven by customer-designed products and services." More specifically, Lee and Thornton believe that central to being "agile" is the ability of a company to be "agile in design" [16]. The ability to understand a product and what is critical therein is key to the success of a company and its ability to become more agile.

As a matter of fact, AM is clearly not always the best choice as a production model in manufacturing. AM can be beneficial as a dynamic organizational model in chaotic markets where customers are promptly attracted by innovations because its objective is to allocate, move, and/or remove production resources whenever necessary. By contrast, it is less effective in stable markets in which a more traditional, static production model may be more appropriate because production resources can, for all practical purposes, be stably allocated over time.

Some enabling technologies, such as the standards for product exchange; concurrent engineering; virtual manufacturing; component-based hierarchical shop floor control systems; and information and communication infrastructures, etc., are critical to successfully achieving AM [11]. In general, innovation in information technology and in organizational models should shorten the path toward agility. In today's business environments, information technology and organization are two strictly intertwined concepts, and it is not easy to assess, *a priori*, which of them is the engine driving the other to evolve. One thing is certain: Information technology is expected to provide a means to allow the effective flow of information within an efficient organization and outside the organization boundaries.

Agile production requires horizontal integration among plants, suppliers, and customers. In this respect, supply chain management provides radical improvements in lead times, reducing stock levels, and the risk of defects, while improving the quality of goods and services. Furthermore, vertical integration, too, is a prerequisite within the manufacturing process. All typical business functions must be related to the supply chain. Information technology should support an organizationally consistent model, allowing the complete representation of the manufacturing process inside the company along with its interaction with customers and suppliers.

In order to guarantee this integration today, many firms, ranging from process control/automation contractors to large management consulting groups, are forced to tackle plant system integration; this entails difficult challenges requiring a sound knowledge of engineering, information technology, and production operations. The success of these integrators depends on their ability to adopt the right implementation tools to the needs of the system being integrated.

The most common strategy of plant system integrators seems to lie in a bottom-up, engineering-centered approach that focuses on specific technical integration capabilities rather than on a common architectural plan for business and plant systems based on appropriate models. To users, this approach often appears inadequate to provide effective solutions, mainly due to wanting information technology. Because of this, manufacturing software applications are usually delivered as fragmented and hard-to-administer point solutions; moreover, these applications become technically complicated and less flexible because of the proprietary software packages and interfaces used in their development. As a consequence, software engineers and plant system integrators must usually establish relationships with plant engineers so as to define plant production aspects (the drawback of which is that the manufacturing entity's business decision makers do not perceive the strategic value in what they deem to be "technical" projects); they must also develop relationships with managers and decision makers in order to discuss business and market features (because plant engineers do not grasp the strategic value in "marketing" projects).

Modern AM must be able to exploit an underpinning technology that supports the coordinated modeling of production and business processes and aims at their continuous improvement following the principle of manufacturing goods and services "from the cradle to the grave." Using this technology, the user should be able to analyze, deploy, and administer his view of the manufacturing system according to his interest and knowledge. In this scenario he would receive support in the supervision and testing of the relations among various objects (i.e., machinery, controls, processes, software applications, etc.) and be able to implement, execute, update, and maintain a certain degree of automation and intelligence among the processes dealing with these objects. For example, the user would be able to design, implement, execute, update, and maintain the network characterizing the launch of a new production order — the set-up of machines; downloading of parameters; analysis of data quality; forecasting of results; etc.

The overriding objective of this technology should not be to implement and execute each of these functions, but to support the user in controlling and coordinating these activities through a set of small, distributed, autonomous, configurable, intelligent, and communicating systems designed to satisfy their specific goals while globally achieving peak performance of the overall manufacturing enterprise's production and business system. Agent technology seems to satisfy this need.

AGENTS AND MULTIAGENT SYSTEMS

We are living in an information society, and the availability of software technologies dealing with highly distributed information, such as Internet-based technologies, has been subject to remarkable growth. As an aside, software agents designed to solve information overloads are probably the fastest growing current of information technology.

The concept of "agent" stems from the lexicon of distributed artificial intelligence (DAI) popular in the 1970s. Research on agents and multiagent systems (MASs) has since flourished, embarking on a myriad of paths and touching on numerous applications, to which the plethora of possible definitions and classifications for agents and MASs attest.

Maes [17] provided the following definition of an agent: "a computational system which is long lived, has goals, sensors and effectors, decides autonomously which actions to take in the current situation to maximize progress toward its (time varying) goals." The same author went further to define a software agent as a "particular type of agent, inhabiting computers and networks, assisting users with computer based-tasks." On the Internet, for example, agents are programs that can gather information or perform some other services without an immediate user presence.

Figure 1.1 shows a possible representation of a generic software agent, highlighting its nature as a self-contained component able to live and communicate in an environment, i.e., an information world, by means of sensors and actuators specific for information management. A single software agent perceives or communicates with other software entities (like services and databases), which do not act to pursue a specific objective, but only to satisfy requests. Agents are active components. An agent is also able to communicate with the physical world, receiving data from devices (e.g., measures or alarms) and sending control signals or sending and receiving messages from users. A software agent can also work in computer networks by receiving and sending data, messages, and signals to possible remote destinations.

Wooldridge and Jennings [18] identified three different classes of agents:

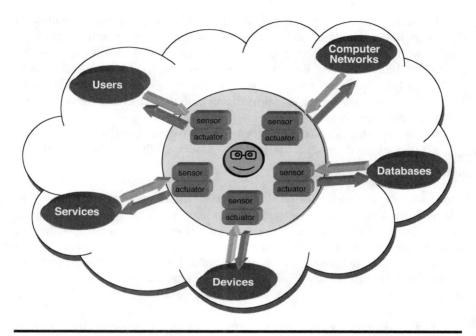

Figure 1.1 Software Agent and Its Information-Based Environment

- Agents that execute straightforward tasks based on prespecified rules and assumptions
- Agents that execute a well-defined task at a user's request
- Agents that volunteer information or services to a user whenever it is deemed appropriate, without being explicitly asked to do so

The main characteristics of these agents are [18]:

- Autonomy, because agents should be able to perform most of their tasks without the direct intervention of humans and should have a degree of control over their own actions and their own internal state
- Social ability, because agents should be able to interact with other software agents and humans
- Responsiveness, inasmuch as agents should perceive their environment and respond in a timely fashion to changes occurring there
- Proactiveness, because when responding to their environment, agents should exhibit opportunistic, goal-directed behavior and take the initiative when appropriate

- Adaptability, meaning that the agent should be able to modify its behavior over time in response to changing environmental conditions and to an enhanced knowledge about its problem-solving role
- Mobility, because the agent should possess the ability to change its physical location to improve its problem-solving capacity
- Veracity, i.e., the assumption that an agent will not knowingly communicate false information
- Rationality, because an agent should be expected to act in order to achieve its goals and not to prevent its goals from being achieved without good cause

Nwana [19] gave yet another perspective on the agent paradigm. The main characteristics an agent should exhibit have been identified in a set of three attributes: autonomy, cooperation, and learning. Although truly smart agents possessing all three characteristics do not yet exist, a more complex range of agent typologies has been defined on the grounds of the previously mentioned characters as well as other characteristics [19]:

- Collaborative agents emphasize autonomy and cooperation to perform tasks by communicating and possibly negotiating with other agents to reach mutual agreements; these are used to solve distributed problems in which a large centralized solution is impractical.
- Interface agents are autonomous and utilize learning to perform tasks for their users; the inspiration for this class of agents is a personal assistant that collaborates with the user.
- Mobile agents are computational processes capable of moving throughout a network, interacting with foreign hosts, gathering information on behalf of the user, and returning to the user after performing their assigned duties.
- Information agents are tools used to help manage the tremendous amount of information available through networks such as the World Wide Web and the Internet.
- Reactive agents represent a special category of agents that do not possess internal, symbolic models of their environments, but instead act or respond according to stimuli arising from the environments in which they are embedded.
- Hybrid agents are particular in that they combine two or more agent philosophies within a single agent.
- A heterogeneous agent system refers to a collection of two or more agents with different agent architectures.

As a matter of fact, as also observed by Imam and Kodratoff [20], there is consensus that a univocal definition of "agent" does not exist because an accurate description depends on the operating objective as well as problem context. For example, the classification used by Woolridge and Jennings [18] is oriented toward the definition of "how" an agent is stimulated to start its actions, whereas the typology identified by Nwana [19] seems to place more weight on agent tasks and applications and on the way agents cooperate or compete to reach their objectives.

Table 1.1 attempts to summarize the main types of agents indicated in the literature, associating them with their more relevant attributes (reported in the columns). The three classes of agents identified in Woolridge and Jennings [18] have been denoted here, respectively, as *rule driven*, *user driven*, and *volunteer*, depending on the cause of their actions, and as *task oriented* and *information/service oriented*, depending on their purposes. The columns under the headings of autonomy, veracity, and rationality are always checked, since all these aspects could be considered as primary features that any agent must possess.

More generally, the variety and wealth of definitions and their related research always emphasize two main views of agents — namely, as stand-alone systems and as multiagent societies. In the first view, the individual features of the agent (e.g., the agent–environment behavior, its proactiveness, the definitions of intelligence and of adaptability, etc.) and its relationship with the human user (e.g., the services that an agent can offer) are generally the focus for development. In this respect, the agent is considered a "decision making artifact" [21] made by a designer; its actions depend on rational reasoning and interests that attempt to reproduce those of the agent's user or owner.

The second view stems from research on complex problems inherent in several sectors [22] and emphasizes the architectural organizations of agents as relational, communication, and network systems, enhancing the distributed nature of multiagent societies. This research has spawned two different methodological approaches: distributed problem solving (DPS) and MAS. The first decomposes, usually in a top-down order, a complex problem into a hierarchy of subproblems whose solution is delegated to a distributed agent system. Agents, interacting and cooperating among themselves, are able to achieve a global solution to the problem in adequate time and at a reasonable cost [23]. The second approach, which by contrast is usually bottom up, focuses on societies of strongly autonomous entities attempting to accomplish local goals and not necessarily cooperating among themselves, whose interaction and coordination depend on their own convenience. In this case, a global solution to the problem is not always guaranteed. The distinction between these two

Table 1.1 Summary of Agent Types and Characteristics

Classes	Autonomy	Social Ability	Responsiveness	Proactiveness	Adaptability	Mobility	Veracity	Rationality	Learning
						Attributes			
Rule driven; task oriented	✓		✓	✓			✓	✓	Sometimes
User driven; task oriented	Partially						✓	✓	
Volunteer information/ service oriented	✓		✓	✓			✓	✓	Sometimes
Collaborative	✓	✓	✓				✓	✓	Sometimes
Interface	✓		✓	✓			✓	✓	Sometimes
Mobile	✓		✓		✓	✓	✓	✓	Sometimes
Information	✓		✓				✓	✓	Sometimes
Reactive	✓		✓				✓	✓	

approaches is not always so clear-cut in more complex problems such as organizational and production systems.

DPS and MAS can be summarized as flexible networks of problem solvers that can tackle problems that cannot be solved using the capabilities and knowledge of the individual solver [24]. The term "MAS" will be used from here on to define any system including two or more agents. In general, an MAS can accommodate many different agent types, each performing specialized functions. In addition to interacting (even partially) with the basic components of their environment in an MAS (as shown in Figure 1.2), some agents can communicate with each other in order to cooperate or to provide some service or information. For instance, specialized "shop floor watcher" agents are in charge of promptly reacting to unexpected events in the physical world perceived by their sensors and, at the same time, reporting the fact to high-level, knowledge-based agents that elaborate a recovery strategy, ultimately forwarding summary information to a human manager by means of interface agents.

Agents operating within an MAS may seem less intelligent than individual agents. However, thanks to their ability to integrate according to specific communication and decision protocols (such as, for example, the contract net protocol [25]), they can solve or support the solution of even more complex problems. In manufacturing, traditional decision and software modeling approaches (for example, Petri nets [26, 27] or Unified Modeling Language (UML) [28–30]) can prove useful to the design and implementation of a MAS.

FROM AGENT TECHNOLOGY TO MANUFACTURING PRACTICE

Agents are not the panacea for industrial software [31, 32]. In addition, although their application can be undoubtedly justified, agents may seem more like a philosophical concept — good for research exercises rather than a useful tool for practical, real-life, manufacturing information technology solutions. Generally speaking, the use of agent technology is justified when the applications are modular, decentralized, changeable, ill-structured, and complex [33]. From a software engineering standpoint, agents may generalize the concept of object programming in which some object methods implement a model of proactive behavior. From a decision and control support perspective, agents may represent the elementary entities introduced to realize decentralized decision policies.

Parunak [31, 32] defined the possible industrial applications of agents from the product life cycle point of view, analyzing three specific areas in which agents have been used effectively: product design, process operation at the planning and scheduling level, and process operation at

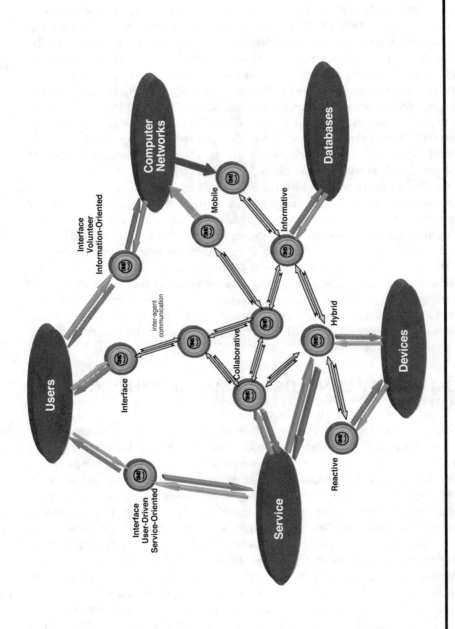

Figure 1.2   Example of a Generic MAS

the lower level of real-time equipment control. According to Parunak's analysis, agents should work as a background tool, allowing engineers in industry to concentrate on the requirements of the problem at hand (that is, the processes related to the product they manufacture). In general, agent technology should support manufacturing engineers with information management. In fact, agents are destined to enjoy increasingly wider use in manufacturing information management systems.

Agent technology also seems to satisfy the fundamental requirements that a modern manufacturing system will need, namely [15]:

- Enterprise integration
- Distributed organization
- Interoperability
- Open and dynamic structure; cooperation
- Integration of humans with software and hardware
- Agility
- Scalability
- Fault tolerance

To satisfy these prerequisites, data are nowadays seen to be as much of an enterprise resource as are raw materials; in this respect, information is manufactured according to a proper workflow, which continuously acquires, integrates, and distributes data related to production, business, and market processes.

Furthermore, the spread of Web-based e-commerce has amplified this practice, with consequent noticeable impact on recent MAS architecture [34]. Agents can actively support this information management. Specifically, in manufacturing, agents can be applied to enhance existing manufacturing information systems (MISs) for purposes of a twofold integration (Figure 1.3):

- Vertically, in order to integrate plant and business processes
- Horizontally, in order to implement automatic information procedures according to a proper market-oriented, distributed workflow management within and outside the enterprise's boundaries

Workflow management is often regarded as a mysterious "add on" in an information system and, in certain cases (such as MISs), as an illusion supplied by expensive and complex software (hardly affordable to small to medium-sized enterprises [SMEs]) or by simpler tools that cannot provide an effective interface in the distributed world of manufacturing. Multifunctional high-confidence distributed databases, which allow different teams to work together in unison with an exact knowledge of the company's

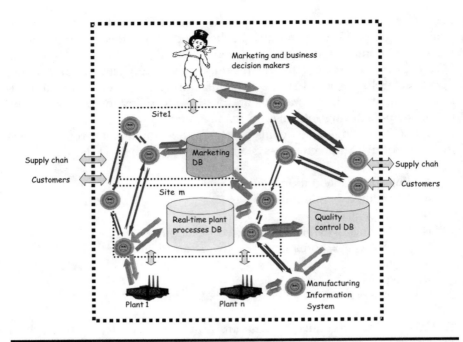

Figure 1.3   Agents Can Support the Manufacturing Information System in Vertical Integration between Plants and Business Processes, and in Horizontal Integration for Internal Workflow Management toward Customer Relationship and Supply Chain Management

current state of production and resources, as well as resistant wide area networks implemented to improve communication, are still a software utopia for many enterprises. The main capabilities requested of these databases are the integration of design into order and scheduling tasks so as to calculate delivery dates; the allocation of resources; and the work schedules of different teams at various sites throughout the product life cycle, that is, from design through production to delivery.

Reliable information processing should support manufacturing industries in their efforts to perform these tasks efficiently. The information to be processed is generally varied in structure and purpose. For example, the following classes of data are most commonly processed in modern manufacturing:

- Procurement management data (warehouse management, order management)
- Production data (shop floor-level process management)
- Aggregation and association of data for general management purposes (forecasting, strategic decision making)

The users (processes, agents, human users) accessing or managing these data may be geographically distributed. However, the integration of these data into a common format is generally necessary so that the enterprise has reliable, real-time information describing its current state and can readily use it for its various production, market, and business activities. A modern approach to manufacturing data management must be based on an MIS that is able to communicate and acquire real-time knowledge on the state of production. The same kind of demand is on the rise for quality control purposes as well.

Taking into account performance, which is required for information retrieval, and given the geographic extension and varied deployment of modern enterprises — not to mention the need to distribute the information to several locations — a centralized information system is not an appropriate solution. Furthermore, given the global nature of modern manufacturing, the information must now flow through systems with very different characteristics, each of which must guarantee process performance. These systems should be integrated taking into account all of the production and business layers of the enterprise, with production processes logically interconnected to the greatest extent possible. As a result, the MIS tends to evolve as a set of integrated systems and intra-/Internet communications among processes in real time. This distributed scenario permits the definition of a more flexible enterprise, enhancing enterprise clustering for possible contractions or expansion. The installation of systems can take place over time to suit the enterprise and each cluster can be fully justified economically and functionally before proceeding with the next phase of the plan. This increases productivity, decreases the time for positive return on investment, and greatly reduces the risk associated with the implementation of a large information system project.

The scalability of an information system in manufacturing is even more important when other facets are considered. For example, in many countries, such as in European Union member states, SMEs, long deemed to be the enterprise model *par excellence* on which the future economy should be based, need small yet effective information systems to be competitive. Scalability is also justified from a hardware point of view. The explosive growth of the Internet has encouraged the use of intranets in the enterprise. PC technology now enables large-scale implementation of inexpensive distributed networks in manufacturing enterprises in which only expensive workstations were previously used. Distributing tasks to autonomous entities and integrating the results are the strategic guiding principles underpinning the implementation of AM, and an MAS provides the technology to do it.

The introduction of concepts like IMS, HMS, and AM serve to satisfy the increasing need for more adaptability and flexibility in the production industry in response to market changes. More and more frequently, scheduling reactions to dynamic variations can be made only in real time. In addition, just in time (JIT) production [35–37], in which an item is not made or purchased until it is needed by the customer or as input to the production process, is generally a strategy that suffers from brusque changes in logistic chain events. Thus, planning and operational decisions tend to be increasingly distributed among the various decision entities, instead of being centralized, to provide more direct and flexible responses to emerging decision problems locally [38]. Integrated information systems such as enterprise resource planning (ERP), computer networks, and, in general, new communication technologies represent a suitable framework for new methodologies and models. The dimension of the distribution of complexity in an MAS can vary according to design specifications and to the agent decision capability in the manufacturing and logistic processes. For example, an agent could simply select e-mails or search for market information on the Internet. An agent could monitor shop floor data to detect anomalies. An agent could be responsible for specific production job orders, checking production whenever specific facts are verified (such as the arrival of a raw material or the state of a production batch) and autonomously deciding which production or marketing strategy to adopt.

Decision capability seems to be more important than intelligence for an agent supporting manufacturing processes and information management. Intelligent agents that are able to learn from previous experiences are, while admittedly desirable, not a must. On the other hand, an agent's knowledge and behavior should be simple to modify. Specifically, the capability to make decisions and the proactiveness to enact them are a must for agents supporting manufacturing. In this respect, agents would represent a heuristic approach especially suited for the highly dynamic and complex problems of modern manufacturing.

In their survey, Shen and Norrie [15] discussed a number of motivations that justify an agent-based approach to manufacturing systems. They observed that centralized hierarchical production management and control systems seem to be unable to react promptly and at low costs to changes in the market or supply chain. This is particularly true for integrated information systems such as ERP systems, which are extremely difficult and expensive to modify or upgrade and usually do not seem tolerant of local failures. Centralized hierarchical systems are also difficult to integrate with different brands of external software packages (without certification of compliance, their reliability cannot normally be guaranteed) and, by definition, do not allow the flexible distribution of information. This is a critical aspect because it represents a hurdle to the effective integration of different companies; this only aggravates the unwillingness they may harbor to share strategic information with potential competitors or to delegate decisional authority to third parties. Centralized systems have a further drawback in that the amount of information needed to manage the enterprise in its many transformations could become so overwhelming that the control of the production processes would be virtually impossible and the details of data so numerous that their added value for management and decision making would be irrelevant.

Agents in an MAS can decide cooperatively [39] for a joint global objective. In this case, an MAS will seldom provide a solution to a problem better than the one that could be found by a centralized decision system with an equivalent computation power. The MAS approach to decision making is justified whenever distributed decision making is requested — for example, when some data cannot be shared. In complex manufacturing decision problems, an MAS approach can provide an acceptable suboptimal solution faster and at a lower cost by distributing the computation to the different MAS components. In this respect, agents perform similarly to common decisional models (such as Petri nets [26, 27], parallel programming [40], cellular automata [41], etc.) found in practical applications. MAS decisional architectures have the same pros and cons found in the distributed decisional models currently in vogue: flexibility and distribution of computational complexity vs. more complex dynamics or a more difficult guarantee of properties such as stability and deadlock avoidance.

With respect to decisional aspects, Lerman [42, 43] introduced a classification of deliberative agents characterized by a certain degree of decisional and intelligent capabilities and swarm paradigms that are nothing more than an MAS composed of extremely simple agents. In the case of deliberative agents, a centralized control managed by a supervisor agent may be present, while in the second case (swarm), the control is distributed among agents and obtained by their interaction. In both cases, the essential characteristics are: robustness, stability, adaptability, and scalability. Conferring these characteristics is, in general, difficult and the decisional aspects and their relation with MAS dynamics can be studied deeply at microscopic levels for each single agent or at macroscopic levels (for the specific features of the whole system's general behavior) [42, 43].

The concepts of macroscopic and microscopic views should call to mind the concept of holon introduced at the beginning of this chapter. At this point, it is important to review the differences between a holon and an agent. A holon is made up of software and/or hardware [13–15], which can be made of other holons. An agent is a system with autonomous behavior with decisional and computational capabilities. It does not include other agents. In general, except for some types of reactive agents, an agent is a software entity. The software component of a holon can be implemented by one or more agents, which can generally access or control the hardware devices of the holon. For example, a transport system based on automated guided vehicles (AGVs) may be modeled by an AGV holon that is responsible for the decisions about vehicle movements and by a holon that is the actual AGV. The same system modeled by agents would be based on a single agent deciding on vehicle movements with the capability to communicate control signals to the AGV that, in this case, is just the physical extension of the agent actuators.

In conclusion, although it should be evident that agents could be useful tools to support manufacturing, it is not quite so clear how to implement agent technology. The purpose of this book is to illustrate how it can be put into practice.

BOOK MOTIVATIONS AND PURPOSES

Manufacturing is undoubtedly one of the most promising fields for the application of agent and MAS technology. Concepts like AM, FMS, and HMS seem to point to the fact that agents and MASs are "the right technology" for manufacturing. The literature is replete with high-level scientific publications on agent and MAS technology, some of which focus on manufacturing and control systems with actual case studies.

This book attempts to highlight the practical facets entailed in the application of agent and MAS technology to manufacturing and control systems. The goal is to show, pragmatically, "how to" add an MAS layer to an existing MIS in order to enhance production by making it more "agile" at a low cost and without the usual revolution involved in implementing a new information system.

The book is organized in seven chapters. After this opening chapter, the second chapter focuses on MAS architecture concepts and on "how to" design an MAS for manufacturing. Here, the basic techniques of high-level MAS design will be explained, introducing the decomposition of a generic planning, control, or scheduling problem into basic entities that can be modeled as autonomous agents. Chapter 3 delves further into the aspects and techniques described in the previous chapter, investigating more deeply planning, scheduling, and control problems. Here, original techniques developed by the authors and established techniques of other researchers will be dealt with in detail. Chapter 4 focuses on the simulation of an MAS and explains "how to" simulate an MAS, for example, to support MAS design in manufacturing. In agent research, simulation is commonly used as a way to validate the model being designed. This chapter provides useful hints on the use of simulation in designing agent-based systems.

Chapter 5 will tackle MAS implementation and "how to" make agents work in and collaborate with the manufacturing enterprise's production processes. The chapter thoroughly reviews the use of object-oriented techniques in designing these systems; the appropriate information and telematics tools to use; and the interagent communication standards with which to comply. Finally, Chapter 6 will give a detailed benchmarking of the most relevant real-world applications for agent-based manufacturing and Chapter 7 will discuss the future challenges for MAS technology in manufacturing, presenting what has yet to be done in agent-based manufacturing and what must be refined. These last two chapters seek to provide researchers with clear suggestions and insight on the most promising — but not fully explored — trends in agent-based research.

To enhance the practical aspects of the book, apart from Chapter 6, each chapter closes with a case study related to some of the MAS methodologies and techniques described. The case study involves a fictional manufacturing SME: PS-Bikes (any resemblance to actual or forthcoming people [with exception of the authors' family names], organizations, or products is purely coincidental). PS-Bikes, as the name suggests, produces bikes and has two production facilities (Plant 1 and Plant 2) in separate geographic locations. PS-Bikes wishes to enhance its production, making it more agile in response to a decision to open its business directly to the consumer through e-commerce. In the last parts of Chapter 2 through Chapter 5, the authors will report their experience in taking PS-Bikes along the path toward agile manufacturing and MAS technology design and implementation.

This book is meant to appeal to a broad public. Management and technical or research professionals, as well as manufacturing enterprise decision makers, who require crucial and practical information dealing with the application of agent technology in agile manufacturing should find this book "must" reading. Although no specific knowledge is required a priori in order to understand the contents of the book, readers should be familiar with the basics of manufacturing processes and management, decision support methodologies, and information technology.

REFERENCES

1. Koste, L.L. and Malhotra, M.K., A theoretical framework for analyzing the dimensions of manufacturing flexibility, *J. Operations Manage.*, 18, 75, 1999.
2. Tan, B., Agile manufacturing and management of variability, *Int. Trans. Operational Res.*, 5, 375, 1998.
3. Koestler, A., *The Ghost in the Machine*, Arkana Books, London, 1967.
4. Valckenaers, P., Editorial of the special issue on holonic manufacturing systems, *Computers Ind.*, 46, 233, 2001.
5. Leeuwen, E.H. and Norrie, D.H., Intelligent manufacturing: holons and holarchies, *Manuf. Eng.*, 76, 86, 1997.
6. Browne, J., Dubois, D., Rathmill, K., Sethi, S., and Stecke, K.E., Classification of flexible manufacturing systems, *FMS Mag.*, 114, 1984.
7. Rembold, U., Blume, C., and Dillmann, R., *Computer-Integrated Manufacturing Technologies and Systems*, Marcel Dekker, New York, 1985.

8. Harrington, J., *Computer Integrated Manufacturing*, reprint, Krieger Publishing Co., Malabar, FL, 1993.
9. Burgess, T.F., Making the leap to agility, *Int. J. Operations Prod. Manage.*, 14, 23, 1994.
10. Goldman, S.L, Nagel, R.N., and Preiss, K., *Agile Competitors and Virtual Organizations*, Van Nostrand Reinhold, New York, 1995.
11. Cho, H., Jung, M., and Kim, M., Enabling technologies of agile manufacturing and its related activities in Korea, *Computers Ind. Eng.*, 30, 323, 1996.
12. Gunasekaran, A., Agile manufacturing: enablers and an implementation framework, *Int. J. Prod. Res.*, 36, 1223, 1998.
13. Christensen, J.H., Holonic manufacturing systems: initial architecture and standards directions, in *Proc. 1st Eur. Conf. Holonic Manuf. Syst.*, Hanover, Germany, 1994.
14. Deen, S.M., A cooperation framework for holonic interactions in manufacturing, in *Proc. 2nd Int. Working Conf. Cooperating Knowledge Based Syst. (CKBS'94)*, DAKE Centre, Keele University, U.K., 1994.
15. Shen, W. and Norrie, D.H., Agent-based systems for intelligent manufacturing: a state-of-the-art survey, *Knowledge Inf. Syst.*, 1, 129, 1999.
16. Lee, D. and Thornton, A.C., Enhanced key characteristics identification methodology for agile design, Agile Manufacturing Forum, March 1996 Boston, MA, 1996.
17. Maes, P., General tutorial on software agents, available at http://pattie.www.media.mit.edu, 1997.
18. Wooldridge, M. and Jennings, N.R., Intelligent agents: theory and practice, *Knowledge Eng. Rev.*, 10, 115, 1995.
19. Nwana, H.S., Software agents: an overview, *Knowledge Eng. Rev.*, 11, 205, 1996.
20. Imam, I.F. and Kodratoff, Y., Intelligent adaptive agents: a highlight of the field and the AAAI-96 Workshop, *Artif. Intelligence Mag.*, 18, 75, 1997.
21. Boutilier, C., Shoham, Y., and Wellman, M.P., Economic principles of multiagent systems, *Artif. Intelligence*, 94, 1, 1997.
22. Durfee, E.H., Lesser, V.R., and Corkill, D.D., Coherent cooperation among communicating problem solvers, *IEEE Trans. Computers*, C-36, 1275, 1987.
23. Bond, A.H. and Gasser, L. (Eds.), *Readings in Distributed Artificial Intelligence*, Morgan-Kaufman Publishers, San Mateo, CA, 1988.
24. Sycara, K.P., Multiagent systems, *Artif. Intelligence Mag.*, 10, 79, 1998.
25. Smith, R.G., The contract net protocol: high-level communication and control in a distributed problem solver, I*EEE Trans. Computers*, C29, 1104, 1980.
26. Zha, X.F., An object-oriented knowledge based Petri net approach to intelligent integration of design and assembly planning, *Artif. Intelligence Eng.*, 14, 83, 2000.
27. Lin, F. and Norrie, D.H., Schema based conversation modeling for agent-oriented manufacturing systems, *Computers Ind.*, 46, 259, 2001.
28. Bauer, B., Müller, J.P., and Odell, J., Agent, UML: a formalism for specifying multiagent software systems, *Int. J. Software Eng. Knowledge Eng.*, 11, 207, 2001.
29. Odell, H.J.J., Parunak, H.V.D., and Bauer, B., Representing agent interaction protocols in UML, in *Proc. 1st Int. Workshop Agent-Oriented Software Eng.*, Ciancarini P. and Wooldridge P., Eds, Limerick, Ireland, Springer, Berlin, 121, 2001.

30. Bergenti, F. and Poggi, A., Exploiting UML in the design of multi-agent systems, *in Proc. ESAW Workshop at ECAI 2000*, Berlin, Germany, 96, 2000.
31. Parunak, H.V.D., Industrial and practical applications of DAI, in *Multiagent Systems: a Modern Approach to Distributed Artificial Intelligence*, Weiss, G., Ed., MIT Press, Cambridge, MA, 1999.
32. Parunak, H.V.D., A practictioner's review of industrial agent applications, *Autonomous Agents Multi-Agent Syst.*, 3, 389, 2000.
33. Jennings, N., Applying agent technology, plenary presentation at PAAM'96, London, 22–24 April 1996.
34. Ulieru, M., Norrie, D., Kremer, R., and Shen, W. A multi-resolution collaborative architecture for Web-centric global manufacturing, *Inf. Sci.*, 127, 3,2000.
35. Monden, Y., *Toyota Production System,* 3rd ed., Kluwer Academic Publishers, Derdrecht, The Netherlands, 1998.
36. Monden, Y., A simulation analysis of the japanes just-in-time technique (with Kanbans) for a multiline, multistage production system, *Decision Sci.*, 15, 445, 1984.
37. Ohno, T., *Toyota Production System: Beyond Large-Scale Production*, Productivity Press, Portland, OR, 1988.
38. Sousa, P. and Ramos, C., A distributed architecture and negotiation protocol for scheduling in manufacturing systems, *Computers Ind.*, 38, 103, 1999.
39. Gou, L., Luh, P.B., and Kyoya, Y., Holonic manufacturing scheduling: architecture, cooperation mechanism, and implementation, *Computers Ind.*, 37, 213, 1998.
40. Agha, G.A. and Kim, W., Actors: a unifying model for parallel and distributed computing, *Artif. Intelligence Eng.*, 14, 83, 2000.
41. Ligtenberg, A., Bregt, A.K., and van Lammeren, R., Multi-actor-based land use modelling: spatial planning using agents, *Landscape Urban Plann.*, 56, 21, 2001.
42. Lerman, K. and Galstyan, A., A general methodology for mathematical analysis of multi-agent systems, usc information sciences technical report ISI-TR-529, 2001, available at http://www.isi.edu/~lerman/papers/papers.html.
43. Lerman, K., Design and mathematical analysis of agent-based systems, in *Lecture Notes in Artificial Intelligence* (LNAI 1871), Springer Verlag, 222, 2001.

2

ISSUES IN DESIGNING
AGENT-BASED
MANUFACTURING SYSTEMS

This chapter addresses the most important practical issues in designing agent-based manufacturing systems. After an introduction to the specific part of the information system in which software agents can find their most suitable applications in manufacturing, typical classes of manufacturing problems that can be faced by agent technology are considered and defined, mostly focusing on planning, scheduling, and control problem definitions. Thereafter, fundamental techniques to break down a generic planning, scheduling, or control problem into basic entities, which can be modeled as autonomous agents, are introduced. Finally, some details are reported about the first meeting with the PS-Bikes managers, who wish to be acquainted with the possibilities of supporting their manufacturing system with agent technology and the practical issues entailed in designing agents and MASs.

INTRODUCTION

Manufacturing, defined as *the action of making or fabricating from material or producing by labor* [1], has been a characteristic human activity for thousands of years. Thus, manufacturing has obviously been performed without any software, in particular any software agent, support for many years. At the same time, a wealth of information is also "manufactured" in parallel to production activities, and this information is increasingly needed to allow agile manufacturing. In this respect, throughout this book, agents are regarded as systems that can enhance manufacturing, support-

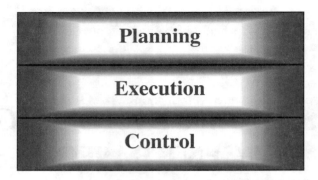

Figure 2.1   Traditional Three-Tiered Model of Manufacturing Information

ing the process in tackling the many problems characteristic of the current information society boom; in doing so, problems typical of manufacturing, namely, planning, scheduling, and control, are also addressed.

This assumption entails the concept that the agent-based manufacturing system — as it is defined throughout this book — is built on an existing system already supported by one or more modules and/or layers of an information system focusing on different aspects (from the shop floor to management), and in which agent societies can find their role in supporting, perhaps even solving, specific tasks neither fully nor efficiently handled at present. Only when their role becomes so relevant that some parts of the current legacy systems are no longer needed can agents be considered an alternative to existing software solutions. In other words, agents here are never advocated as an "*ab initio*" design solution for a manufacturing system; rather, their role is to integrate existing functionalities, to support new ones and, only in case their work is proven satisfactory, to substitute the modules of the legacy system that are obsolete.

It therefore becomes important to introduce the environment in which an agent should act, that is the information system of a manufacturing enterprise, before we describe just how agents can be integrated. Traditionally, computer software systems in manufacturing management have been designed on three layers [2] (Figure 2.1):

- The planning layer, including material requirements planning (MRP); its later development, manufacturing resource planning (MRPII); and its latest evolution, enterprise resources planning (ERP)
- The execution layer, including the manufacturing execution system (MES), which bridges the gap between the planning and the control systems by using on-line information to manage the current application of manufacturing resources: people, equipment, and inventory

■ The control layer, including hardware and software device control systems (DCSs), programmable logic controllers (PLCs), supervisory control and data acquisition (SCADA) systems, etc.

This three-tiered model is already an innovation compared to a more traditional "plan and control" production system, due to the introduction of the MES layer. This layer was introduced in the 1990s [2] in response to the need to integrate many disparate pieces of information, such as statistical process control; the tracking of work order, time, and employee attendance; quality control reports; and accomplished production. MES can be handled as the operational level of a manufacturing system; it has been defined in the MESA-11 standard [2], and in the more recent ANSI/ISA-95 standard, as manufacturing operations and control level [3]. Computer software systems developed to achieve this integration are still generally called MESs. An MES is expected to provide the following core functions (see McClellan [4] for more details):

■ Planning of the system interface, that is, passing information to/from each neighboring layer
■ Work order management, that is, the automatic or manual uptake of information about what and how much is to be produced
■ Workstation management, which implements the direction of the work order plan and the logical configuration of workstations; the planning, scheduling, and loading of each operational workstation are performed here
■ Inventory tracking and management, which entail the processing, storage, and maintenance of the details of each lot or unit of inventory (which is, in turn, anything needed for production)
■ Material movement management, which prompts the movement of a specific inventory unit to the workstation
■ Data collection, which functions as a clearinghouse for and translator of all information needed and/or generated within the production facility
■ Exception management, which is the ability to respond to unanticipated events that affect the production plan

In addition, an MES is expected to provide the following support functions:

■ Maintenance management, providing historical, current, and planned maintenance events
■ Time and attendance control — for example, badge-scanning systems

- Statistical process control focusing on the continuous monitoring of a process rather than the inspection of finished product
- Quality assurance/ISO 9000
- Data processing/performance analysis
- Documentation/product data management
- Genealogy/product tracking
- Supplier management (specifically in the case of outsourcing and just-in-time inventory management)

In fact, one of the reasons underpinning the introduction of MES was the need to answer to the increasing demand for agility in manufacturing [5]. On the other hand, although the MES layer can improve the vertical integration of a traditional "plan and control" production system, as described in the previous chapter, agile manufacturing also requires a deep horizontal integration, for example, among customers and suppliers, supported mainly by Internet/intranet technologies. This need has led to another evolution of the information systems found in manufacturing management that is exemplified in more recent models. For example, the availability of the supply-chain operations reference model (SCOR) [6] (Figure 2.2) recently had a major impact on business system planning. This model breaks down supply chain management (SCM) into four main processes:

- PLAN: related to typical production planning activities, such as the definition of the master production schedule
- SOURCE: related to activities typically associated with the management of providers and the inventory

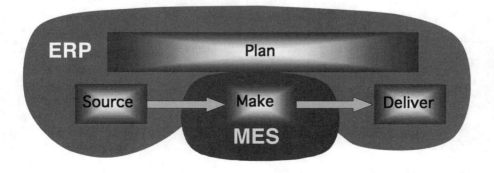

Figure 2.2 Supply Chain Operations Reference (SCOR) Model (Redrawn from Supply Chain Council, SCOR model, http://www.supply-chain.org, 1996.)

- DELIVER: related to the management of customers and the distribution of products
- MAKE: related to production processes

With respect to the SCOR model, ERP products can address the requirements of the PLAN, SOURCE, and DELIVER processes, while MES, jointly with control layer components, is the primary component of the MAKE process. To this extent, the combined pressure of ERP implementation and supply chain strategies has begun to force manufacturing companies to focus increasingly on their plant systems. Most large and mid-size companies are deploying ERP systems but, as they attempt to extend such systems to make them cover not only planning but also execution and control functions characterizing the plant activities, it becomes evident that a gap in operation functionality exists. This leads manufacturing organizations to look for a new approach to plant operation management. The ability to create an effective link between plant resources and production management, together with tight closed-loop integration among plant, suppliers, and customers, has become increasingly necessary. The REPAC (READY, EXECUTE, PROCESS, ANALYZE, and COORDINATE) model [7] incarnates this idea. In particular, REPAC (Figure 2.3) addresses all of the processes required to operate the plant and coordinates all factory activities with the remaining part of the logistic flow.

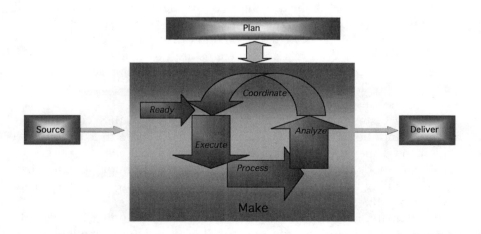

Figure 2.3 The REPAC (READY, EXECUTE, PROCESS, ANALYZE, and COORDINATE) Model (Redrawn from AMR Research, Inc., The AMR Report on Manufacturing, 1998.)

- READY: Readies production for the introduction of new product lines, or for engineering modifications to old ones. In fact, this process allows the definition of the most suitable plant configuration for the manufacture of new or modified products. It also administers product and process improvements, including corrective actions for detected noncompliance.
- EXECUTE: Executes an optimized production plan to ensure that products comply with their specifications. It also provides PROCESS with the suitable machinery configuration (set-up) needed to make a specific product. The majority of historical MES functions are included in this step.
- PROCESS: Includes all the features needed to produce the product physically. It also provides a means to automate and control the process.
- ANALYZE: Allows supervisory personnel to analyze meaningful data from all sources and, thus, to evaluate production performance, product quality, process capability, and regulatory compliance. Data are also made available to the ERP and supply chain management system, as well as to customers and suppliers for analysis purposes.
- COORDINATE: Coordinates plant operations with the enterprise and supply chain. It defines the optimized sequence of plant activities to meet the production requirements defined by demand management.

The REPAC model highlights the need for tight integration among all these business processes. Manufacturing is not a "static" process; it is the result of a continuous flow of actions and information inside the company. Data are as much a company's resource as are raw materials, machinery, and labor. For world-class manufacturing, the entire manufacturing process must be highly flexible and responsive to the frequently changing requirements of the supply chain.

Table 2.1 summarizes the relations between the MES core/support functions and the REPAC model. Actually, some of these functions can be implemented in more than one REPAC process. As it can be observed, the REPAC model also includes product and process improvements, i.e., typical planning activities, in the READY process that are not traditionally covered by MES functions.

In summary, it can be roughly assessed that the planning layer is able to manage manufacturing processes "off-line" in a long to medium term; the control layer is able to manage the production process with "hard real-time" capabilities in a medium to short term, while the MES layer is able to manage many functionalities guaranteeing flexibility and integration between planning and control with "soft real-time" capabilities. Some

Table 2.1 REPAC vs. MES Core and Support Functions

REPAC process	MES core function	MES support function
READY		Maintenance management
EXECUTE	Workstation management Inventory tracking and management Material movement management	Process data/performance analysis
PROCESS	Exception management Inventory tracking and management Material movement management	Process data/performance analysis
ANALYZE	Data collection	Statistical process control Quality assurance/ISO 9000 Process data/performance analysis Documentation/product data management Genealogy/product traceability
COORDINATE	Planning system interface Work order management Workstation management	Maintenance management Time and attendance Supplier management (outsourcing and just-in-time inventory management)

Controls definition and MES to controls data flow possibilities, White Paper No 3 ed., 1995.

functionalities can be replicated in the different layers, but the approach is generally different due to the different time constraints.

MES seems to provide a fitting environment for agent living, and the problems that software agents can respond to here can generally be traced to traditional planning, scheduling, and control problems, taking into account soft real-time constraints. The SCOR SOURCE and DELIVER processes of a supply chain, due to the recent impact of Internet/intranet technologies, may also be a friendly environment for software agents. In addition, Internet/intranet by definition requires on-line interactions, whenever soft-real time functions are needed, some supply chain management functions are also likely to be managed by MES. See, for example, the supplier management support function defined by McClellan [4] that is taken into account as an MES support function when special management is required, such as outsourcing and just-in-time inventory management.

In fact, the future requirements of manufacturing information systems seem to support the idea that a layered structure [8] will no longer be adopted in favor of a new structure to be integrated as one monolithic, although distributed, structure. The layered architecture will not work in true agile manufacturing, and no additional layer will solve the problem. The typical refinement carried out in this architecture — that is, planning production at a medium term interval (1 month, 1 week); taking into account rough and aggregate information about production capacity and operational constraints; executing daily plant production with the problems of the real world; and controlling the workstation performing jobs, does not fulfill the requirements of agile manufacturing. This last task is to react promptly to continuous external stimuli, integrating production with decisions related to suppliers and customers. Agile manufacturing nullifies the benefits coming from integrating production management into layers because, for example, the planning layer works over too long an interval and with inefficient information.

Although this scenario may be considered apocalyptic for managers of current legacy systems, it is the authors' opinion that agent technology can also be viewed as a way to introduce the required integration and new functionalities to their enterprise gradually.

In this chapter, issues and techniques to design an MAS, mainly for purposes of supporting a legacy system of a manufacturing enterprise rather than substituting it, are described. Because, unlike other settings, the MAS design phase in a manufacturing system requires a deeper analysis of when/why/where to introduce an MAS, this chapter will focus more on the analytical aspects of design, leaving matters related to software architectures to the fifth chapter. After a review of the main issues and existing techniques available to design an MAS, a practical approach applied to the fictitious firm, PS-Bikes, is presented.

AGENT SYSTEM ENGINEERING

Before any attempt can be made to implement agent societies effectively in a manufacturing system, an analysis of the industrial life cycle is pivotal. In fact, this cycle poses restrictions and constraints on the development of an agent-based system that are not present in most research environments [9]. Techniques to discover where and how an MAS can be profitably fitted to the industrial life cycle are generally necessary. Once the problem or problems are identified, an MAS design phase, which is more oriented toward implementation, starts; here important aspects, e.g., the MAS architecture, the specific capabilities of each single agent, and the interagent communication, are defined. In this respect, important frameworks, such as the one proposed by the Foundation for Intelligent

Physical Agents (FIPA) [10], have already begun to gather consensus as agent technology standards.

As a matter of fact, from many standpoints, although MASs are already taught as an academic discipline, their design is still practiced as an art. Thus, methodologies and techniques are needed to bridge the gap typically separating "theory" from "practice," aiming to limit practical drawbacks and hurdles that are surely encountered in real-world applications of agent technology. Müller [11] is one of the first authors who clearly expressed the need for a systematic design of MASs. In his review, Müller cited three main types of recipes to construct MASs: the use of formal specification frameworks, like the task-based approach [12]; the declarative representation approach [13]; and architecture-oriented approaches like the beliefs, desire, and intentions (BDI) agents [14]. However, because at that time rationale and criteria underpinning the decision for a special design or technique did not exist, more pragmatic and general approaches were advanced; one of these was the bottom-up method of agent/world/interoperability/coordination [11], which incorporated ideas from the structure of orgafunctional design and the design of simulation systems.

Later on, Sycara [15] claimed two technical hurdles to the extensive use of MASs: first, a lack of a proven methodology enabling designers to structure applications clearly as MASs; second, no general case industrial-strength toolkits flexible enough to specify the numerous characteristics of agents. Since then, although many efforts have been geared toward software implementations and numerous approaches toward standards for MAS design have been proposed, an effective standard-driven methodology does not yet exist.

A more recent survey of agent-oriented software engineering [16] presents an updated state-of-the-art of design methodologies. In this survey, agent-oriented software engineering is defined and divided into high-level methodologies, which are addressed more to an analysis activity and design methods that lead up to an implementation activity (Table 2.2). In this respect, Wooldridge and Ciancarini [17] affirmed that the primary approach to develop methodologies for MASs entails the adaptation of those developed for object-oriented analysis and design, bearing in mind attendant limitations such as the problem that object-oriented methodologies simply do not allow capturing many agent systems features, like proactivity. In fact, a comprehensive and rigorous methodology for the development of multiagent systems is lacking, and system developers have paid little attention to requirements specification and the analysis process [18].

In this chapter, high-level methodologies are taken into account as the first issues of the design phase, leaving the description of proper design methods like UML-related methodologies such as PASSI (process for agents

Table 2.2   Classification of Agent-Oriented Software Engineering

Agent-oriented software engineering	
High-level methodologies	*Implementation-oriented design methodologies*
GAIA	Uml oriented: AIP, AUML, PASSI ...
MaSE	Design patterns
AOR	Components
	Graph theory

Source: Tveit, A., A survey of agent-oriented software engineering, in *Proc. 1st NTNU CSGS Conf.* (http://www.csgsc.org), http://www.jfipa.org/publications/AgentOrientedSoftwareEngineering/, 2001.

societies specification and implementation [19]), which are more related to a real agent implementation, to the fifth chapter. Therefore, considering the aims of this book, design methodologies have been grouped into three main interrelated approaches (Figure 2.4):

- Problem-oriented MAS design, guided mainly by user requirements
- Architecture-oriented MAS design, guided mainly by software aspects
- Process-oriented MAS design, guided mainly by the constraints of the environment (for example, a plant workstation) where the MAS will be put into operation

Figure 2.4   High-Level Design Methodologies

Table 2.3 Types of Problems

Major type	Type of problem	Generic conclusion
Synthesis	Modeling	Behavioral model
	Design	Structure of elements
	Planning/reconstruction	Sequence of actions
Modification	Assignment (scheduling, configuration)	Distribution/assignments
Analysis	Prediction	Discrepancy state

Source: Reprinted from *CommonKADS Library for Expertise Modelling*, Breuker, J. and Van de Velde, W. (Eds.), with permission from IOS Press.

These approaches are obviously complementary and should always be taken into account in the design of any MAS, at microlevel (agent structure) as well as macrolevel (agent society and organization structure) detail. The problem architecture process terminology is mainly related to the feature or property on which design is based; however, as with any software project, the design approach of the final product should ideally implement all three aspects.

Problem-Oriented MAS Design

The path toward the introduction of an MAS into a manufacturing company can be oriented by the identification of the reasons for which the system is needed. Usually, an MAS is adopted to solve existing problems or to enhance management aspects of the company by adding intelligence to the existing information system.

At a higher level, the identification of a problem is a process of knowledge engineering. Breuker [20] identified this process as a sequence of steps. The output of the first step, the "problem identification stage," which is obtained from spontaneous, ill-defined problems, stems from a discrepancy between a current state and a norm state. This output results in the identification of a conflict between the desirable and the real behavior. The second step, the "problem definition" stage, takes this identified conflict in input to produce as output some abstract solution, which is usually related to some known problem type or a sequence of problem types. For example, Table 2.3 shows a suite of problem types [20]. Once types have been identified and the problems have become well-defined ones, problem-solving methods allow transforming them into tasks. When the knowledge engineering process is aimed at designing an MAS rather than a knowledge-based system, each task should be implemented in one or more agents, possibly reusing agents. High-level design methodologies oriented toward manufacturing problems should follow a more or less similar knowledge engineering approach.

Among the high-level MAS design methodologies, the GAIA and the MaSE approaches seem to be the most promising. The first [21, 22] is a general methodology that, while supporting the microlevel and macrolevel of agent development, requires that interagent relationships (organization) and agent abilities be static at run-time. The GAIA *analysis* process starts by finding the *roles* in the system and continues by modeling *interactions* between the roles found. Roles are based on four attributes: responsibilities, permissions, activities, and protocols. *Responsibilities* are of two types: *liveness properties*, i.e., the role should add something good to the system, and *safety properties*, which prevent and disallow problems in the system. *Permissions* define what the role is allowed to do: specifically, which information it is allowed to access. *Activities* are tasks that a role performs without interacting with other roles. *Protocols* are the specific patterns of interaction. In the GAIA *design* process, the first step is to map roles into *agent types* and then to create the right number of *agent instances* of each type. The second step is to determine the *services model* needed to fulfill a role in one or several agents, and the final step is to create the *acquaintance model* for the representation of communication between the agents.

Wood and DeLoach [23] proposed the multiagent systems engineering (MaSE) methodology. MaSE improves on GAIA by providing support for automatic code creation through a specific tool. The goal of MaSE is to lead the designer from the initial system specification through to the implemented agent system. The MaSE methodology is divided into seven sequential steps:

1. *Capturing goals*: The initial system specification is transformed into a structured hierarchy of system goals.
2. *Applying use cases*: Use cases and sequence diagrams based on the initial system specification are created. Use cases present the logical interaction paths between various roles.
3. *Refining roles*: Roles that are responsible for the goals defined in the first step are created. Each goal is represented by one role, but a set of related goals may map to one role. Together with the roles, a set of tasks is created that defines how to solve goals related to the role. Tasks are defined as state diagrams.
4. *Creating agent classes*: Roles are mapped to agent classes in an agent class diagram.
5. *Constructing conversations*: A coordination protocol is defined in the form of state diagrams that define the conversation state for interacting agents.

6. *Assembling agent classes*: The internal functionality of agent classes is defined. Selected functionality is based on five different types of agent architectures: belief–desire–intention (BDI); reactive; planning; knowledge based; and user defined.

7. *System design*: Actual agent instances based on the agent classes are created; the final result is presented in a deployment diagram.

In the search for a design methodology more attuned to real manufacturing problems, in a recent survey Parunak [24] focused on methodologies to create industrial agent systems. Among others, DaimlerChrysler's solution [25] is the most oriented to the analytic aspects of MAS design, aiming to provide a task-oriented approach to agent design. The DaimlerChrysler approach shares many similarities in terms of role identification and inter-role interaction with the GAIA and MaSE methodologies cited earlier, but the search for a proper definition of tasks, if stressed, takes it to an implementation design phase.

Concerning its reusability and its specific application to the information system of a manufacturing company, problem-oriented MAS design seems to be most effective for two main situations:

- Complex distributed decisions (e.g., some scheduling problems), in different formulations and in different settings, which are characteristics of the information layer in which they act; these agents will be referred to as *synthetic social agents*
- Management activities in which an MAS can automate certain business operations, such as workflow management or customer relationship management (CRM); these agents will be referred to as *business component agents*

Both of these issues will be exemplified in the case study at the end of this chapter.

Architecture-Oriented MAS Design

The design of any information system relies heavily on the architecture. Many authors accept agent and MAS architecture as the first step of design, using one of its instantiations to describe the problem-solving process. In this respect, Müller [11] listed the following motivations:

- Guidelining: An architecture specification of the agent is by definition a valuable general guideline for the MAS design, as well as for the implementation of the application.

- Structuring: An architecture specification of the agent generally provides the description of the system's modules and layers, which are the classes of operational knowledge necessary to design and build the MAS.
- Re-use: An architecture is generally related to an implicit execution model which avoids programming from scratch.
- Standards: An architecture is generally related to a set of standards allowing many advantages, among which is the awareness of the possibility to communicate with other systems. In this respect, predefined application-independent mechanisms are usually directly available to the developers as a standard procedure.
- Predictability: The MAS behavior, through the basic patterns of interactions of the instantiated agents, can be predicted up to a certain level.
- Genealogy: An architecture generally allows strategic and functional extensions according to the evolutions requested by the environment in which an MAS acts.

At microlevel, the simplest agent architecture (Figure 2.5) consists of a module for communication, a knowledge base, and a reasoning module [11]. In this elementary, three-layered agent architecture, many issues can be specified. The specification of the communication module should be related to two main considerations: the channel through which agents can communicate and the content of their dialogues. Because an agent is a software application that lives in a computer network, it should be compliant with or referred to computer network standards. For example, an agent and the related MAS should be clearly referred to an ISO/OSI computer network architecture, specifying several aspects stretching from a physical point of view (e.g., its bandwidth requirements) all the way to its software properties at a low level (e.g., whether it uses TCP-IP in a client–server or message-passing architecture) and a higher level (the language syntax specification, the need for security/authentication, etc.).

The specification of the knowledge base (KB) should be related to the encapsulation of the agent's know-how in a certain domain. In this respect, the KB of a manufacturing system should be interpreted broadly to include if–then rules; parameters resulting from the computation of a back propagation standard algorithm of an artificial neural network; parameters related to the optimal configuration of a scheduling problem, etc.

Reasoning is related to the definition of the KB. A typical example is the inference engine, which is able to fire rules of the KB according to a specific algorithm. When the reasoning module, on its own, is also able to modify the KB content, adding rules or modifying parameters, the

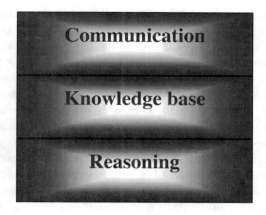

Figure 2.5 Layered Architecture-Oriented Design of an Agent

reasoning module is also intelligent — that is, it is able to learn from experience. Thus, even if traditional manufacturing does not always require this feature, the reasoning module clearly represents the most fascinating aspect of research on agent and MAS.

The macrolevel MAS architecture is also an important design aspect. According to Parunak [9], the design process of an MAS, especially when developed for an industrial application, is heavily architecture oriented and, like other software and information systems, its refinement is an iterative process. Parunak and colleagues [26] proposed an approach, defined as "synthetic ecosystems," predicated on four main stages: conceptual analysis; role-playing; computer simulation; and implementation design. Even though not all these stages are provided with a supporting analysis, their definitions allow giving concrete contributions to MAS design.

Conceptual analysis gives an initial vision of what the system, as a whole, will do. A broad set of constraints should be taken into account here (e.g., interface, performance, operating, life-cycle, economic, political); however, conceptual analysis places key focus on the definition of the desired system behavior, thus breaking down the system in a top-down approach to identify agents. This identification can be performed following a linguistic case analysis in which specific nouns (such as "unit process," "resource," "manager," "part," "customer," "supplier"), if used as defined by the desired system behavior, can guide, though not as a finished system design, this breakdown into candidate agents. Candidate agents should be validated according to some general principles such as thing vs. function; smallness in size; decentralization; diversity and generalization, etc. Once agents are identified, the definition follows of their individual behavior and of the classes of messages they can exchange, including aspects of cooperation and of organization. Aspects such as the definition of the "stimulus" and the related "response" in the agent behavior; concurrent planning and execution; local communication; and information sharing are also defined. Role-playing is subsequently performed to test the system.

This exercise focuses mainly on the architecture of the agent system, supported by speech acts and related graphs. The actual test can be undertaken once the subsystems to be role-played are selected, scripts to guide role-playing activities are written, and agents are assigned to people. Role-play actions are recorded on cards and include five pieces of information: the identity of the sending agent; the identity of the receiving agent; the time at which the card is sent; the identity of the agent whose card stimulated this one; and the time that the card stimulating this one was sent. This information allows reconstructing the thread of conversation among the agents. Enhanced Dooley graphs [27] provide a useful tool to analyze conversations in agent-based systems. Computer simulation, focusing on the dynamics of the MAS behavior, which is often supported by nonlinear mathematical analysis, and focus of implementation design on platform and tools are the two subsequent stages.

Shen et al. [28] described a generic collaborative agent system architecture (CASA) specific for intelligent manufacturing systems. In this architecture, some important and original aspects are the need to enhance the aspects of agent cooperation and the need to create order in the MAS (for example, registering agent services and locations by the use of yellow page agents).

In conclusion, several MAS architectures have been proposed in the literature because this is the traditional approach to design information systems. Architecture MAS design should always be preferred whenever information system aspects are the predominant component.

Process–Oriented MAS Design

Another approach to the design process, which is always related to the system architecture and which is probably attractive for control engineers, is that which starts from the definition of the time constraints posed by the different processes in the manufacturing system. In fact, agent-based manufacturing systems may need to interact with the external world or with the plant and related human operator-driven activities. For these reasons, they should show some kind of real-time behavior in order to be able to react to asynchronous signals or to respect time constraints.

Most of the application cases in which users may need time determinism within their software applications demand that the software architecture have the possibility to achieve real-time tasks. To cope with such

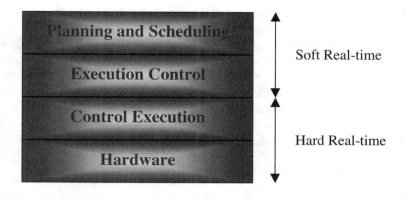

Figure 2.6   Example of Real-Time Holonic Control Architecture (From Zhang, X. et al., *Inf. Sci.*, 127, 23, 2000.)

problems, many specific real-time operating systems exist by guaranteeing real-time reliability, but also cause the program to be executed in a nonstandard run-time environment. In other applications, time constraints may not be so rigid. This leads to a classification into hard and soft real-time applications, always keeping time determinism as a mandatory requirement. In manufacturing applications, where real time is strongly needed and different layers are managed by different software and hardware configurations, both aspects (soft and hard real time) are present: in particular, low-level plant control tasks are performed by dedicated programmable logic controllers in a hard real-time environment (e.g., level control in a tank), while supervision and monitoring operations are performed at higher production levels and with different needs. In this case, thanks to the presence of graphical user interfaces, the interaction with humans, and network connections that are slower with respect to some field-busses, the real-time framework can be much broader.

In this respect, partially on the basis of a previous work by Shen et al. [28] and related to a traditional view of the manufacturing information systems, Zhang et al. [29] proposed a four-layered architecture for real-time holonic control that is summarized in Figure 2.6. The upper layer handles high-level manufacturing production requirements, from schedule generation to supply chain management. The execution control layer is composed of "holonic units" responsible for control application management, while control execution is introduced to "control what is being executed" and supports requirements such as fault detection and recovery. The two upper layers have soft real-time requirements, while the remaining lower layers entail hard real-time requirements. Implementation characteristics, such as the extent of distribution and the real-time capabilities of the adopted operating system, are thus already defined at the design stage.

Focusing on processes, and in order to guarantee that the proposed real-time agents can deal with the variety of hardware and software that may be encountered in a flexible manufacturing plant, platform independency becomes a must. The Java platform [30], whose claim is "write once, run everywhere," is an ideal candidate solution for this purpose. It can be argued that Java may not be the best appropriate language for real-time programming; on the other hand, it offers all of the other features needed for the development of software agents, including an international standard for agent implementation and communication [10].

Given the general characteristics that a real-time system should offer, Java does not seem to fit at all for real-time programming, primarily because Java is a platform-independent language and a real-time scheduling of Java threads cannot be guaranteed if no *a priori* knowledge about the scheduling characteristics of the operating system is available. Moreover, Java has an automatic garbage collector whose influence has been immediately criticized [31]. Nevertheless, because Java naturally supports threads, it appears to be suited for some kinds of real-time programming. In fact, real-time applications must be written as a series of separate component programs that can execute concurrently in a multithreading organization. Indeed, every Java thread is a complete program capable of independent execution, sharing its memory with others and always keeping separate addresses to ensure rapid context switching. Moreover, since it first appeared, Java has been presented as a language for facilitating the development of embedded systems software, and most embedded computer systems must deal with real-time constraints.

In order to provide Java with real-time characteristics, a first solution is to modify Java Virtual Machine, as Perc [32] did, which can be considered a real-time dialect of Java. An alternative solution would seek to develop specific real-time programming facilities for the handling of process scheduling and real-time features that allow respecting the time constraints of (soft) real-time applications. Some degree of real-time reliability with Java is also possible, given that Java programming is performed in a special environment. Moreover, many embedded systems applications increasingly require a Web connection [33]; Java is the best language for Web programming because it has a special capability for the development of distributed applications.

In conclusion, process-oriented design is probably the preferred methodology of control engineers. This approach is justifiable whenever the control layer of the manufacturing information system has some aspect to be enhanced.

So, Which MAS Design?

No standard design methodology supporting the analysis aspects of manufacturing production and business processes exists at present. Many approaches provide a good reference point from which to start; on the other hand, there is no unique recipe for the design of an agent or an MAS. In general, current engineering approaches to MAS design focus on:

- Supply chain modeling and its related industrial life-cycle problems (problem-oriented approach)
- The information system of the manufacturing company and related workflow activities (architecture-oriented approach);
- Shop-floor modeling and its related control problems (process-oriented approach)

However, this distinction may not be so apparent and the support of an architecture to be considered as a reference model is usually required.

In any case, before starting MAS design, an evaluation must be made as to whether an MAS is the best solution to solve the problem at hand. A proper analysis comparing MAS solutions to other possibilities will assess the compliance of manufacturing IT with user expectations, chief of which are [34]: support to the manufacturing strategy (continuous improvement, job shop, flow line, etc.); support to investment in the supply chain; the possibility of implementation on existing automation; the possibility of installing it quickly (weeks), incrementally with a minimum of staff time; the possibility to be quickly rolled out to additional plants; and the possibility to upgrade to new revisions without redoing integration and configuration.

ISSUES IN DESIGNING AGENT-BASED MANUFACTURING SYSTEMS AT PS-BIKES

A Short Description of PS-Bikes

PS-Bikes is a small enterprise that produces *make-to-order* bicycles. PS-Bikes is situated in the south of a European country and provides bikes to retailers all over Europe, but does not currently sell bikes to individual customers. The company is made up of two main departments: an administrative office, located in a downtown area with major communication and commercial infrastructures that expedite relations with partners and clients and the promotion of new products, and a production facility located a few kilometers away in the countryside.

The production process involves the production of the frames of the bicycles and several assembly phases; in addition, test and quality control phases are performed. The bicycle components, such as wheels, tires, and gears, are ordered to partner companies that manufacture them according to the PS-Bikes designs. The bicycle frames are produced in the PS-Bikes plant from raw materials. Three models of bicycles are manufactured: children's bikes, mountain bikes, and racing bikes. Two color qualities are available: five different pastels and two different metal colors. Figure 2.7 shows the plant layout and summarizes the process flows. Raw material (iron tubes) and finished bike components are stored in two different warehouses. Iron tubes are standard lengths long enough to manufacture the largest frame size and undergo quality control according to a defined protocol. Finished bike components arrive with quality control already certified.

After quality control, iron tubes arrive at the frame areas where three laser cutters are programmed to prepare the tubes for frames, according to specified orders. Once cut, the tubes are assembled into single kits (one per bike) and stored in a buffer area. Kits of cut tubes are then sent to welding, a process consisting of two phases: welding and heat treatment, performed by two different machines: a welder and an oven, respectively. The frames are thus assembled and ready to be painted. Two different lines are specialized to paint the frames with a pastel or a metal color. Once painted, the frames are sent to an area for drying. Finished frames are stored with the other finished bike components in the specific warehouse. All finished components are then brought to an assembly area where the desired product is assembled. Bikes are finally stored in a finished product warehouse, waiting for their delivery to the customers.

Although small, PS-Bikes possesses a well-organized information system, which is broken down into three layers: planning, MES, and control. The company's information system stores data in a relational database management system (RDBMS), whose tables are distributed in two servers, one in the administration department and the other in the plant. The firm is quite satisfied with this information system; however, management feels that some modifications are necessary to answer to changes in production. This need is likely to involve the MES layer more. A brief description of the functionalities of the PS-Bikes information system, with more emphasis on the MES layer, follows.

Planning processes are performed in the administration department. Planning activities include typical supply chain processes, including CRM, warehouse management, provider management, and the definition of the master production schedule (MPS). PS-Bikes' MPS includes the list of orders to be filled. This list of orders is computed each week and passed to PS-Bikes' MES.

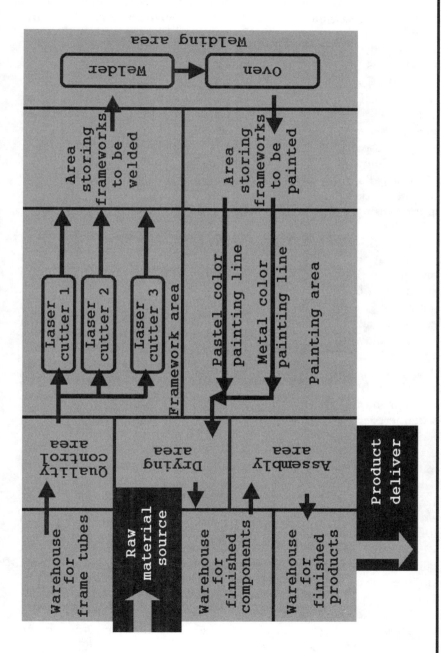

Figure 2.7 Layout of the PS-Bikes Existing Plant and Related Process Flows

PS-Bikes' MES aims to satisfy the production plans of the administration department. MES processes, satisfying the functionalities defined in the introduction of this chapter, are performed in PS-Bikes' plant. The MES can accomplish these functionalities, producing and processing information following the production flow. Practical examples of MES processes performed in PS-Bikes can be given following an example of a typical production day.

The first function that PS-Bikes' MES is able to perform is the planning system interface. For a given week, a list of orders included in the MPS is received. This list is a quite complex ensemble of structured information related to customer orders. These orders are defined in batches, each usually containing more than one bike to be produced with the same characteristics. PS-Bikes' MES is able to add information in relation to real current production and to modify the MPS, e.g., to split one order into more than one or to unite orders in one single order for purposes of efficiency.

Among the functions performed by PS-Bikes' MES is the scheduling of the workstations. All the possible plant operations (such as cutting a race bike frame, painting a frame with a pastel color, drying a frame, etc.) are coded and defined in a specific table. Likewise, workstations (such as laser cutter 1, oven 1, etc.) are coded in another table. Relationships between workstations and operations, i.e., the information about which workstations allow which operation, are also stored. Each day, and more than once a day, PS-Bikes' MES defines the routing of each frame to be manufactured; in other words, it defines the steps of production, fitting the code of each product to the workstation/operation relationship and adding other information related, for example, to the processing time.

To define each routing and, specifically, to define the order of batch processing, PS-Bikes' MES has a procedure that sorts MPS orders according to the earliest due date at the beginning of production. In addition, in each buffer area, a specific scheduling algorithm to establish the order of each single product defines the sequence of processing. For scheduling, the MES uses simple deterministic algorithms [35], which cannot take into account accidental problems, such as, for example, the failure of an operation, the tardiness in completing a process, etc.

PS-Bikes' MES also performs inventory management, a function that is generally executed by the planning layer; here again, however, the MES accomplishes some functionalities that are related to what is needed in real-time execution. For example, while the planning layer has a global knowledge of the processes to feed the warehouse from suppliers, PS-Bikes' MES is able to track any raw material, from its position in the warehouse to its movement in the plant and on to its assembly in the finished bike. Functionalities in the inventory management are thus related to the update of tables containing information about positioning and origin/destination in the movement of raw materials. This is particularly useful because two automatic storage/retrieval systems (AS/RSs) are used in the warehouse, one for frames and the other for the additional bike components; in some areas automated guided vehicles (AGV) are used. The history of these movements, together with data collected from the monitoring of production processes, enables another important function: the genealogy/product traceability function.

The control layer is able to handle manufacturing processes mainly through real-time capabilities given by specialized software and hardware devices. For example, one small programmable logic controller (PLC) is responsible for the processes of the three laser cutters.

Does PS-Bikes Need Agents?

PS-Bikes is concerned about the impact that several possible modifications and upgrades might have on their market. In this respect, three major needs are felt:

- They wish to expand their business to single customers by e-commerce, but at the same time they are aware that the introduction of this kind of process would produce a *bullwhip* effect, with a random unpredictable flow of orders quite incompatible with the regularity of current planning.
- They have recently acquired a new plant in the North, and this is an additional source of organizational problems. PS-Bikes' idea is to equip this plant with an MES very similar to the one already in operation in the older southern plant. The problem here is that, although the MES is optimized and tested in the monolithic planning/MES/control architecture, the thought of a doubled MES raises many doubts. For example, in case unpredicted low production in one plant occurs due to failures and to the attendant unexpected maintenance operations, it should be reasonable that, if convenient, part of the production could be taken up by the other plant. However, this inter-MES communication facility, which most likely also involves the planning layer, is currently not implemented and the dynamics of the processes between the two MESs is unclear.
- Finally, the MES also has some maintenance problems. For example, when a third new laser cutter was introduced, the scheduling algorithm had to be modified and tested, taking into account the characteristics of the new model. Each modification introduced in the plant entails some weeks of software development and testing, and even some days of blocked

production. What PS-Bikes' managers have in mind is a wishful ambition for immediate "plug-and-play" software with each new piece of machinery installed; at the same time, they realize that the several scheduling algorithms present in the plant will need to be reformulated when the configuration of the system is modified.

In other words, PS-Bikes wants to be agile.

The introduction of software agents can help the company achieve this goal. The following section describes an attempt to design a specific MAS that satisfies these needs.

From Problems to Agents at PS-Bikes

A team of MAS designers thus sets out to design a proper MAS architecture for PS-Bikes. The three problems cited previously are quite clear, but their resolution is not so evident. The impact of the three problems is also quite different: the first two require an enhancement of the current software architecture, while the third means that the MES architecture must be redefined. In addition, it is clear that these problems are typical of situations compelling the collapse of a layered system architecture like PS-Bikes' into a single monolithic architecture. Agile manufacturing requires the continuous modification of production according to the overall supply chain status, causing long-term planning to lose its significance. However, for user convenience, this collapsing process will be gradually implemented, as a result of the introduction of the MAS into the system.

Two Planning Problems at PS-Bikes

The first problem mentioned at the beginning of this section regards planning and is quite common in modern enterprises. The possibility to enhance CRM using Internet technologies to satisfy each single customer is very appealing, but it clashes with traditional plan/execute/control organization. As a matter of fact, two main problems due to business-to-customer e-commerce may be innate in the planning of PS-Bikes' activities:

- A great amount of information related to small orders containing generally just one bike, with characteristics and features defined by the individual customer
- An unpredictable flow of orders, which creates "bullwhip" effects in production

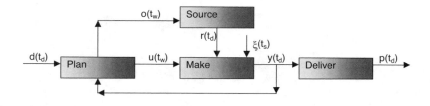

Figure 2.8 Main Variables and Their Relations That Have an Important Impact on Manufacturing Production Processes

Agents here should support efforts aimed at resolving both problems, addressing management activities related to certain operations, such as workflow management and CRM, and introducing intelligence to control the unpredictable aspects of e-commerce.

According to a simplified SCOR model view of PS-Bikes' "make-to-order" management, the following variables (Figure 2.8) have an important impact on the manufacturing production processes:

- $d(t_d)$: the demand of the market collected each day t_d by the administrative office
- $o(t_w)$: the bill of material sent each week t_w to the administrative office carrying out the management of suppliers and of the warehouse
- $u(t_w)$: the set of orders computed in the MPS sent each week t_w to the MES of the plant
- $r(t_d)$: the raw material used each day t_d by the plant to make products
- $\xi(t_s)$: the unpredictable disturbance that always (for example, each second t_s) may affect production performance; examples are engine breakdown and extraordinary maintenance
- $y(t_d)$: related to the finished bikes (and pertinent information) produced each day t_d
- $p(t_d)$: related to the delivery batches scheduled each day t_d and sent to customers

Workflow Management and CRM

The goal here is to support and to automatize a series of functions that can benefit from software implementations. As a consequence, MAS design is strongly architecture oriented.

Workflow systems are among the most well-known technologies addressing this trend. Workflow can be defined as the computerized facilitation or automation — wholly or in part — of a business process. The workflow management system thoroughly defines, manages, and performs "workflows" through the execution of software, whose order of execution is driven by a computer representation of the workflow logic

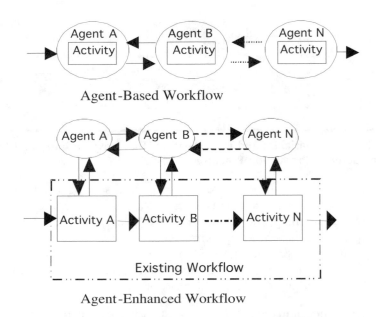

Agent-Based Workflow

Existing Workflow

Agent-Enhanced Workflow

Figure 2.9 Two Basic Agent-Based Architectures for Workflow Management (Redrawn from Montaldo, E. et al., *Inf. Syst. Frontiers*, 5, 195, 2003)

[36]. Given their ability to integrate different ISs, such as those managing plant and business information, workflow management systems constitute one of the major applications of software agents in manufacturing.

Figure 2.9 shows two basic ideas of agent-based architecture applied to workflow management [37]. The first is taken from the advanced decision environment for process tasks (ADEPT) project by British Telecom labs [38]. The ADEPT system consists of multiple software agents that concurrently negotiate an agreement on how resources should be assigned to support a business process. The software agents take full responsibility for business process provisioning, enactment, and compensation, with each agent managing and controlling a given task or set of tasks. The second agent-based architecture, called agent-enhanced workflow, is shown in the lower part of Figure 2.9. Here, agents provide an additional layer to an existing commercial workflow management system. The agent layer takes full responsibility for the provisioning and the compensation phases of business process management [39].

The Manufacturing Agents in a Knowledge-based Environment driven by Internet Technologies (MAKE-IT) project [40–42] is a proposal of heavily architecture-oriented MAS design that should fulfill the requirement of workflow agents in PS-Bikes. MAKE-IT seeks to define and implement "small" software architectures, called MAKE-IT agents, that add functionalities such as workflow management to an existing manufacturing IS. This approach, following the basic agent architectures applied to workflow [37], can be classified as an agent-based architecture applied to an existing workflow management system in order to manage new functionalities, such as the management of CRM in electronic commerce. The environment in which MAKE-IT agents live is an IS in which: 1) data are stored and retrieved from relational database management systems (RDBMS); 2) documents are generated according to events of the manufacturing production process; and 3) information can flow inside the enterprise through a distributed MAKE-IT agent network to and/or from the external Internet world.

The design of a MAKE-IT MAS follows the specification of architecture-oriented design as specified by Müller [11]. Specifically, the MAKE-IT communication model is a message-passing model, based on XML (eXtensible Mark-up Language) [43] tunneled inside the enterprise boundaries within Microsoft Message Queue (MSMQ) [44] software channels, and outside the enterprise within traditional Internet channels. With XML, a computer can easily and unambiguously handle information and can avoid the most common pitfalls, such as lack of extensibility, lack of support for internationalization/localization and platform dependency. The knowledge of a MAKE-IT agent is currently modeled in a rule-based system and reasoning is obtained by an inferential engine (CLIPS [45]). From a functional point of view, the MAKE-IT agent network, where many MAKE-IT agents can work cooperatively, represents a distributed repository of the know-how about specific information processes required by the workflow management, which is a sort of knowledge network. The MAKE-IT architecture also provides agents with a yellow pages directory service, using the active directory (AD). The AD schema contains a formal definition of the contents and structure of AD services, including all attributes, classes, and class properties. The current MAKE-IT version implements a new agent class with the attribute's name, description, and address.

At present, an agent can perform the following basic actions:

■ It can query the RDBMS using the statements SQL select, insert, update, delete (each agent performs a query on a different set of tables).
■ It can send e-mail to customers and to suppliers.
■ It can send messages to other agents within the enterprise.

A practical example of application of the MAKE-IT architecture to PS-Bikes workflow follows.

PS-Bikes' administrative office consists of several business units. The main tasks of the PS-Bikes sales unit pertain to the management of customer orders from retailers and to production scheduling. The main tasks of the purchasing unit are to verify that the raw material warehouse can satisfy production demands at short to medium term (2 weeks) and to send orders to raw material suppliers. The main tasks of the production unit are related to the management of production according to scheduling and to verify and update warehouse information.

PS-Bikes is seeking to expand its marketing efforts on the Web in order to reach single customers. The company's strategy also entails receiving orders from retailers via the Web and modifying the production accordingly, but ideally retaining as much as possible of the company's simple management structure. A Web site is therefore designed to provide access to the orders-from-customers table for the insertion of orders from thousands of probable new customers. According to the MAKE-IT approach, three agencies are added to the PS-Bikes manufacturing IS in parallel to the workflow of the managing units: the purchasing agency; the production agency; and the sales agency. Each of these agencies can manage a specific subworkflow, which, for the sake of brevity, is only outlined here. These agencies are added to the existing layered information architecture, and they autonomously work to integrate and coordinate the single-customer orders coming from the Web site. Specifically:

- The sales agency performs the following actions: 1) it checks whether a new order has been made; 2) it captures information about new orders and controls whether it is possible to ship them (in this case, the agency must be able to send an e-mail to the customer and to update the RDBMS) or not (in this case, the agency must be able to send a message to the production agency); 3) it receives messages from the production agency.
- The production agency performs the following actions: 1) it checks whether it is possible to produce a particular product for a particular order, capturing information from the sales agency, and sends the order to the shop floor; 2) it receives information from the shop floor about the state of production; 3) it sends a message to the sales agency when the product is ready; 4) it updates the RDBMS; 5) it sends a message to the purchasing agency if PS-Bikes is not able to produce the bike specified in the order.
- The purchasing agency aims at inventory reduction, although it does not apply just-in-time management [46]. Management is performed in two ways: 1) it receives a message from the production agency when it is not possible to satisfy the order; 2) it checks whether the inventory is below a fixed threshold. In both cases, a pertinent e-mail is sent to the supplier.

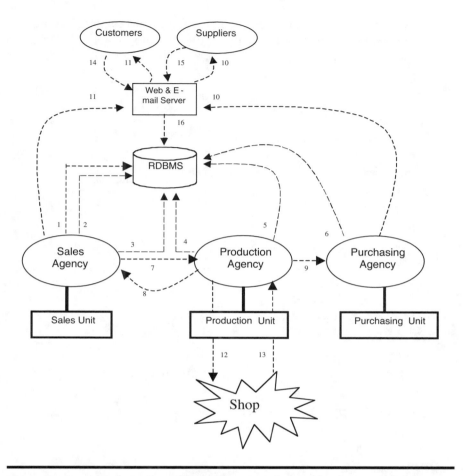

Figure 2.10 Manufacturing Information System integration by MAKE-IT agencies (Redrawn from Montaldo, E. et al., *Inf. Syst. Frontiers*, 5, 195, 2003)

The description of these workflow functionalities is summarized in Figure 2.10, where the numbers are respectively related to:

1. Query of the RDBMS to check if a new order has arrived
2. Query of the RDBMS to check if the order is already in the warehouse
3. Updating of RDBMS
4. Check of the possibility to produce the order
5. Updating of RDBMS
6. Check of whether inventory is under a fixed threshold

7. XML message sent by the sales agency to the production agency when the order is not already in the warehouse:

```
<Order_Production>
<Order_Code>
</Order_Code>
</Order_Production>
```

8. XML message sent by production agency to sales agency about delivery time:

```
<Delivery_Mail>
<Customer>
<Name></Name>
<Surname></Surname>
<E-mail_Address><E-mail Address>
</Customer>
<Product_Code></Product_Code>
<Quantity></Quantity>
<Delivery_Time></Delivery_Time>
</Delivery_Mail>
```

9. XML message sent by the production agency to the purchasing agency when it is not possible to produce the order:

```
<Order_To_Supplier>
<Order_Component></Order_Component>
<Quantity></Quantity>
</Order_To_Supplier>
```

10. E-mail supplier
11. E-mail customer
12. Information to shop floor
13. Information from shop floor
14. Information from customers
15. E-mail from suppliers
16. Information from Web and e-mail server, such as new orders from customers, a new customer registration, or a new delivery from suppliers

These activities are autonomously executed by a simple MAKE-IT architecture, which can be distributed geographically on PS-Bikes's workstations. However, PS-Bikes' employees always have decisional power in any critical situation; for example, if the list of orders of raw materials is

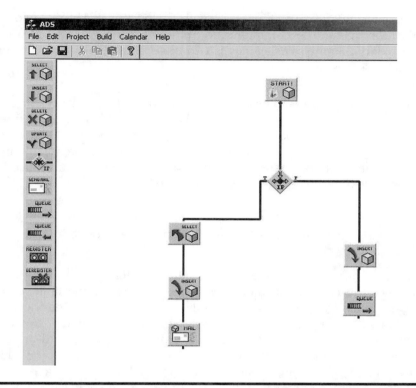

Figure 2.11 Knowledge Modeling Interface Applied to a PS-Bikes MAKE-IT Agent Working in the Sales Agency

prepared and ready to be sent as an e-mail, it can still be viewed and modified by a PS-Bikes' employee before dispatch.

To model the PS-Bikes agent knowledge, some variables are defined: the *agent name*; the *agent address*; the *active directory address* (the address of the primary domain controller in the domain); the *input queue* and the *output queue*; and the name of the *clips file*.

Figure 2.11 represents the knowledge modeling interface applied to a PS-Bikes MAKE-IT agent working in the sales agency. The main task of this agent is to verify whether a new order can be satisfied or not. The simple workflow in Figure 2.11 is related to a query performed in the PS-Bikes RDBMS. If a new order has arrived, the agent must check whether it is already stored in the RDBMS (whether or not that kind of product is available in the warehouse). If the two conditions are satisfied, the agent can update the PS-Bikes RDBMS and can e-mail the customer about delivery time, additional product information, and other general customer relations information. When the product is not available in the warehouse, the agent sends an XML message with a specific order code to the production agent. The communication model for the PS-Bikes MAKE-IT agent working in the sales agency is also quite simple; the agent can send output messages in two queues:

- The public queue *order_production* of orders sent to the sales agency when the product ordered is not available in the warehouse and must be produced
- The private queue *sales_agency_agent_queue:* an internal communication about a current agent's state; this agent can also send e-mail to customers about delivery time

Planning Strategies to Manage the Unpredictable Aspects of e-Commerce

In the previous example, agents are neither intelligent (i.e., they do not learn) nor too complex in their reasoning (i.e., few simple if–then rules represent their knowledge) and, according to the definition introduced in this chapter, can be classified as business component agents. However, thanks to their simple architecture, they are quite efficient and are able to coordinate a wide flow of information. On the other hand, some added knowledge (e.g., grouping of single orders in batches; visualizing products to customers while contracting the offer; or providing some CRM function better) might enhance their role in PS-Bikes.

More specifically, here the MAS design should not be oriented toward implementing automatic business procedures with little knowledge (which a software engine can do more quickly), but toward providing planning strategies to manage the unpredictable aspects of e-commerce. The goal is here to render production stable and to avoid hikes in costs due to the fragmentation of demand. In fact, the unpredictable flow of single orders, each demanding one custom bike to be delivered to a specific address, is an unnerving prospect for PS-Bikes' managers. Thus, they feel the need to break through in B2C e-commerce, but wish to avoid the disruption of their current business and production processes.

Keeping these requirements in mind, one solution might be to cluster customers so that each member has similar characteristics (e.g., delivery address, delivery time, price, type of bike, ...) and so that demand consists of a batch of products comparable to the one of distributors. To make this feasible, single customers should be persuaded and guided to selecting products that allow this clustering. The process can be managed by planning individual customers as traditional bigger clients; if their demand can be oriented in the period of the year when production is lighter, this effect could, paradoxically, result in more stable production.

With reference to Figure 2.8, the variable most related to this problem is the demand $d(t_d)$. This parameter is characterized by some attributes, the most important of which are:

- Number and type of bikes contained in the demand.
- Customer description.
- Price of each bike.
- Delivery time.

Also with reference to Figure 2.8, it is clear that the demand, the very first input of the whole production processes, is the most critical variable and that the ability to identify, predict, and control its dynamics is at the basis of agile manufacturing. The design phase of the requested activities to be added to the current planning layer therefore becomes strongly problem oriented, and specific methods should be introduced to solve the problems. Details about conventional and agent-based planning methods will be discussed in Chapter 3.

A Control Problem at PS-Bikes Southern Plant

In PS-Bikes' southern plant, the control layer is provided with traditional hardware technologies and the control logics are stored on one programmable logic controller (PLC). The PLC is accessible by IEEE 802.3 protocol and, via communication with PCs, also supports TCP/IP; as a result, although not yet implemented, these data are accessible by Internet applications. In addition, production is tracked product by product through the use of several data points in which a mix of technologies (code bars, on-product radio frequency identifiers [RFID], AGVs, etc.) positions products and works in progress in real time, as well as provides the basis for the genealogy functionalities. This setting represents a great opportunity for the introduction of agents.

In addition, some deterministic scheduling algorithms are implemented on workstations. Unfortunately, these algorithms hide some pitfalls: they are deterministic and do not work whenever something is not working, and plant modifications often implicate algorithm modifications. For example, faced with the option to install a new laser cutter, PS-Bikes' staff must answer the questions of whether:

- It is possible to introduce a new laser cutter, as well as any other new workstation or machinery, without disturbing the rest of the scheduling software.

■ It is possible to have scheduling algorithms that are distributed throughout the plant components (products, machinery, network, ...).

Contracting, *cooperating*, and *coordinating* are probably the words that best describe the solutions to the requirements of real-world dynamic scheduling. These aspects will be discussed in Chapter 3.

Scheduling Execution at the New PS-Bikes Northern Plant

Stressing the considerations of the previous subsection, the questions are now:

■ Is it possible to provide the new northern plant with an agent-based holonic MES?
■ Is it possible to introduce agents that allow sharing the MES layers of the two plants as if they were a virtual one?

Actually, new paradigms of virtual manufacturing [47] are geared toward answering these challenges. These aspects, too, will be further analyzed in Chapter 3.

CONCLUSIONS

Designing an agent is still a subjective art and no single methodology provides a proper and failsafe paradigm. High-level methodologies, which have been described in this chapter, are needed to discern whether an MAS may or may not be useful in a manufacturing information system; on the other hand, they do not lend themselves easily to prompt implementation. Implementation-oriented design methodologies, which will be exemplified in Chapter 5, are as their name implies, nearer to actual implementation. However, they are still too far removed from the problems and current characteristics of the working information system.

In this chapter, some problems of PS Bikes, which may compel the introduction of an MAS to its information system, have been introduced. For problems related to workflow management, the MAKE-IT approach to design business component agents, which has been subject of the authors' previous studies, has been introduced and exemplified. MAKE-IT can be also viewed as a strongly architecture-oriented methodology, with features of a high-level methodology, because it allows agent modeling with adequate software tools and implementation-oriented design, because the modeling phase implementation is just an immediate consequence of the design phase. However, it is the authors' opinion that

MAKE-IT could prove most valuable when applied to settings involving repetitive tasks not entailing any special requirements, such as learning or optimization of complex distributed decisions. The example shown of workflow management fulfills these needs. For more complex problems, such as on-line scheduling problems, more extensive arguments and more complex methodologies that allow the design of synthetic social agents should be introduced. These aspects will be discussed in the next chapter.

In conclusion, it is the authors' opinion that agents in manufacturing must be designed in a no-man's land, in which planning and control are not traditional and which might be considered an MES, where software still does not completely or effectively fulfill expectations. In this respect, agile manufacturing requires new planning, scheduling and control methodologies and agents are in the position to cover this role.

REFERENCES

1. Oxford English Dictionary, 2nd ed., 1989.
2. MESA International, Controls definition and MES to controls data flow possibilities, White Paper No 3 ed., 1995.
3. ANSI/ISA-95.00.01-2000, Enterprise-Control System Integration Part 1: Models and Terminology, ANSI/ISA-95.00.02-2001, Enterprise-Control System Integration Part 2: Object Model Attributes, available at http://www.isa.org/.
4. McClellan, M., *Applying Manufacturing Execution Systems*, St. Lucie Press, Boca Raton, FL, 1997.
5. Kelly, T., Electronic data systems, MES in the age of agile manufacturing, a presentation at MESA Roundtable 4, Chicago, IL, September 13, 1995.
6. Supply Chain Council, SCOR model, http://www.supply-chain.org, 1996.
7. AMR Research, Inc., Do we need a new model for plant systems? The AMR Report on Manufacturing, Boston, MA, 1998.
8. McClellan, M., Evolving to the enterprise production system (EPS), presented at National Manufacturing Week, Chicago, IL, March 2000.
9. Parunak, H.V.D., Industrial and practical applications of DAI, in *Multiagent Systems: a Modern Approach to Distributed Artificial Intelligence*, Weiss, G., Ed., MIT Press, Cambridge, MA, 1999.
10. The Foundation for Intelligent Physical Agents, available at http://www.fipa.org/.
11. Müller, H.J., Towards agents system engineering, *Data Knowledge Eng.*, 23, 217, 1997.
12. Brazier, F., Keplicz, B.D., Jennings, N.R., and Treur, J., Formal specification of multi-agent systems: a real-world case, in *Proc. 1st Int. Conf. Multi-Agent Syst. (ICMAS (AAAI, 1995)*, Lesser V., Ed., 25, 1995.
13. Singh, M.P., Huhns, M.N., and Stephens, L.M., Declarative representations of multi-agent systems, *IEEE TKDE*, 5, 721, 1993.
14. Kinney, D., Georgeff, M., and Rao, A., A methodology and modelling technique for systems of BDI agents, in *Proc. 7th Eur. Workshop Modelling Autonomous Agents Multi-Agent World*, LNAI 1038, 1996.
15. Sycara, K.P., Multiagent systems, *Artif. Intelligence Mag.*, 10, 79, 1998.

16. Tveit, A., A survey of agent-oriented software engineering, in *Proc. 1st NTNU CSGS Conf.* (http://www.csgsc.org), http://www.jfipa.org/publications/Agent-OrientedSoftwareEngineering/, 2001.
17. Wooldridge, M. and Ciancarini, P., Agent-oriented software engineering: the state of the art, in *Proc. Agent-Oriented Software Eng. 2000 (AOSE 2000)*, Wooldridge, M. and Ciancarini, P., Eds., 1, 2000.
18. Arazy, O. and Woo, C., Analysis and design of agent-oriented information systems, *Knowledge Eng. Rev.*, 17, 215, 2002.
19. PASSI: a process for agents societies specification and implementation, available at http://www.csai.unipa.it/passi.
20. Breuker, J., A suite of problem types, in *CommonKADS Library for Expertise Modelling*, Breuker, J. and Van de Velde, W., Eds., IOS Press, Amsterdam, NY, 57, 1994.
21. Wooldridge, M.J. and Jennings, N.R., Software engineering with agents: pitfalls and pratfalls, *IEEE Internet Computing*, 3, 20, 1999.
22. Wooldridge, M.J., Jennings, N.R., and Kinny, D., The Gaia methodology for agent-oriented analysis and design, *Autonomous Agents Multi-Agent Syst.*, 3, 285, 2000.
23. Wood, M.F. and DeLoach, S.A., An overview of the multiagent systems engineering methodology, in *Proc. Agent-Oriented Software Eng. 2000 (AOSE 2000)*, 207, 2000.
24. Parunak, H.V.D., A practitioner's review of industrial agent applications. *Autonomous Agents Multi-Agent Syst.*, 3, 389, 2000.
25. Bussmann, S., Agent-oriented programming of manufacturing control tasks, in *Proc. 3rd Int. Conf. Multi Agent Syst. (ICMAS'98)*, IEEE Computer Society, 57, 1998.
26. Parunak, H.V.D., Sauter, J., and Clark, S.J., Toward the specification and design of industrial synthetic ecosystems, in *Proc. 4th Int. Workshop Agent Theories, Architectures Languages (ATAL)*, Springer Verlag, Berlin, 45, 1997.
27. Parunak, H.V.D., Visualizing agent conversations: using enhanced dooley graphs for agent design and analysis, in *Proc. 2nd Int. Conf. Multi-Agent Syst. (ICMAS'96)*, AAAI Press, 275, 1996.
28. Shen, W., Norrie, D.H., and Kremer, R., Developing intelligent manufacturing systems using collaborative agents, in *Proc. 2nd Int. Workshop Intelligent Manuf. Syst.*, Katholieke Universiteit Leuven, 157, 1999
29. Zhang, X., Balasubramanian, S., Brennan, R.W., and Norrie, D.H., Design and implementation of a real-time holonic control system for manufacturing, *Inf. Sci.*, 127, 23, 2000.
30. The source for Java technology, available at http://java.sun.com/.
31. Johnson, R.E., Reducing the latency of a real-time garbage collector, *ACM Lett. Programming Languages Syst.*, 1, 46, 1992.
32. About Perc, available at http://www.newmonics.com/.
33. Perrier, V., Can Java fly? Adapting Java to embedded development, *Embedded Developers J.*, September, 8, 1999.
34. Martin, R., Future manufacturing IT architectures, World Batch Forum, AMR Research, Boston, MA, 2000.
35. French, S., *Sequencing and Scheduling: An Introduction to the Mathematics of the Job-Shop*, Ellis Horwood Ltd., Chichester, England, 1982.

36. Workflow Management Coalition, The Workflow Reference Model, Document Number TC00-1003, January 1995.
37. Odgers, B.R., Thompson, S.G., Shepherdson, J.W., Cui Z., Judge D.W., and O'Brien P.D., Technologies for intelligent workflows: experiences and lessons, in *Proc. Agent-Based Syst. Bus. Context, AAAI Workshop*, 1999.
38. Jennings, N.R. et al. ADEPT: managing business processes using intelligent agents, in *Proc. BCS Expert Syst. 96 Conf. (ISIP Track)*, Cambridge, UK, 5, 1996.
39. Shepherdson, J.W., Thompson, S.G., and Odgers, B.R., Decentralised workflows and software agents, *BT Tech. J.*, 17, 1999.
40. Sacile, R., Paolucci, M., and Boccalatte, A., The MAKE-IT project: manufacturing agents in a knowledge-based environment driven by Internet technologies, in *Proc. Academia/Ind. Working Conf. Res. Challenges IEEE-AIWORK-2000*, Buffalo (NY), IEEE Press, 281, 2000.
41. Montaldo, E., Sacile, R., Coccoli, M., Paolucci, M., and Boccalatte, A., Agent-based enhanced workflow in manufacturing information systems: the MAKE-IT approach, *J. Computing Inf. Technol.*, 10, 1, 2002.
42. Montaldo, E., Sacile, R., and Boccalatte, A., Enhancing workflow management in the manufacturing information system of a small-medium enterprise: an agent-based approach, *Inf. Syst. Frontiers*, 5, 195, 2003.
43. W3 Consortium, Extensible Markup Language (XML), February 10, 1998, available at http://www.w3.org.
44. Microsoft Technet, MS message queue server overview, 1998, available at http://www.microsoft.com/technet/default.asp.
45. Riley, G., Clips: a tool for building expert systems, August 28, 1999, available at http://www.ghg.net/clips/CLIPS.html.
46. Groenvelt, H., The just-in-time system, in *Logistics of Production and Inventory, Handbooks in Operations Research and Management Science*, Graves, S., Kahn, A.R. and Zipkin, P., Eds., North Holland, Amsterdam, 4, 629, 1993.
47. Camarinha–Matos, L.M., Execution system for distributed business processes in a virtual enterprise, *Future Generation Computer Syst.*, 17, 1009, 2001.

3

AGENTS FOR PLANNING, SCHEDULING, AND CONTROL

This chapter focuses on the application of agent-based modeling and agent technology to the three main activities characterizing manufacturing production: planning, scheduling, and control. The objective is to highlight the reasons that make an agent-based approach appropriate for each of these activities. The chapter covers these areas in 10 main sections:

- The first three sections deal with the general role of planning, scheduling, and control in manufacturing, underlining the relevance of an integrated perspective.
- The next three sections recall the necessary concepts relevant to planning, scheduling, and control in manufacturing, providing some key topics about non agent-based approaches, which throughout this chapter will be referred to as "classic" or "conventional" approaches.
- The section titled Agent-Based Applications in Manufacturing Planning and Scheduling summarizes the main features of some of the most outstanding multiagent approaches to planning and scheduling reported in the literature, highlighting the strict integration that agent technology provides for the two phases.
- The next section begins the wrap-up of the chapter by showing how agent-based control in manufacturing can be viewed as a natural extension of planning and scheduling.
- The section titled Planning, Scheduling and Control in the PS-Bikes shows how some previously introduced concepts can be applied to the PS-Bikes case study, together with an original multiagent scheduling approach developed by the authors.
- The final section draws conclusions.

UNIVERSITY OF HERTFORDSHIRE LRC

INTRODUCTION

As discussed in Chapter 2, the role of MASs in manufacturing seems more appropriate in cases in which production activities are affected by dynamic variations or involve complex decisions, and in instances in which the agents are adopted to make systems interoperate, playing a role of intelligent middleware components. In addition, agent technology can be applied by introducing two different classes of agents: *business component agents* and *synthetic social agents*. The main characteristic of the former is to fit exactly into one task or a subset of tasks required to carry out a business process, whereas the latter is usually introduced to tackle complex decisions, decentralizing the decisional capabilities among the different actors of the decision process and modeling this process by means of a social collaboration/competition paradigm. This chapter discusses the application of such modeling considerations from three different stand-points that are respectively associated with the business processes or the complex decisions pertinent to the planning, scheduling, and control in manufacturing.

Among the various application areas that may require one of these activities in a manufacturing enterprise, this chapter will focus on the operational level of a manufacturing system, the one commonly referred to as the execution layer (defined as MES by MESA-11 standard [1, 2] or manufacturing operations and control level as defined in the ANSI/ISA-95 [3] specifications). This choice stems from three considerations:

- The need for applications to confront planning, scheduling, and control problems seems to co-exist in the execution layer.
- Although the role of planning may appear somewhat limited in the execution layer, it seems more effective to compare agent solutions addressing processes of a common layer. In addition, as mentioned in the previous chapter, agile manufacturing will increasingly demand the collapse of the traditional layered structure of a manufacturing system in a monolithic architecture in which long-term planning loses significance.
- The execution layer is that in which things happen and change according to a rapid dynamic; here complex operational decisions must be made, often with only partial information available.

FOCUSING ON PLANNING, SCHEDULING, AND CONTROL ACTIVITIES IN MANUFACTURING

This chapter deals with the application of MASs to planning, scheduling, and control activities in manufacturing systems. The very general aspects relevant to such activities were already introduced in Chapter 2. However, before detailing these applications, the usual meanings of the terms *planning*, *scheduling*, and *control* in manufacturing must be further defined, as must the approaches to face them that are generally proposed by practitioners and scientific communities.

Planning is the activity devoted to defining plans. A plan specifies what an organization wants to achieve in the future. In manufacturing industries, plans may involve strategies (e.g., the acquisition of new market shares or the launch of a new product line) laid out over quite long-term outlooks (quarters or years), as well as decisions about the use of the production facilities for imminent time periods (e.g., the production plan for the next quarter specified on a weekly basis). A plan thus characterizes the behavior of the organization, or part of it. To meet the objectives or requirements specified in a plan, a company (i.e., a manufacturing system — MS) must efficiently use available resources (people, machines, tools, information, and so on) following an appropriate implementation strategy. The success of the company depends on its ability to define or select the most appropriate plan (e.g., the most profitable) and its ability to attain the goals specified therein: namely, on how effective the adopted implementation strategy will prove to be.

Plans need to parallel the company's mission and must first of all be feasible and secondly not easily subject to improvement. For instance, a manufacturing industry seeking to increase its net revenue by producing and selling high-quality products in the current economic scenario would be able to enlarge its customer base or conserve its most profitable customers by satisfying their requests in terms of quality, response time, personalization, assistance, and so on. This could define a mission, which can be pursued following a number of strategies that affect the nature and scope of the company plans. A plan is feasible if it can be realized by exploiting available company resources and within the planned time-frame or, at least, at the moment the plan is defined, if no information is available assessing the contrary. A feasible plan is robust even if its feasibility is guaranteed in the case of limited (small and seamless) variations affecting the scenario that was contemplated when the same plan was conceived. Stating that the plan cannot be easily improved simply means that, even if suboptimal, that plan must be a sound one that does not stray excessively from the optimum.

Many of the concepts introduced so far could be similarly adopted to describe the meaning of the scheduling activities. Planning and scheduling, especially in manufacturing, are very often associated to describe a strategy or a support system. In fact, scheduling aims at defining a particular type of plan, called a schedule, that precisely defines how the available resources must be used to obtain a result (e.g., meet the production requirements for the next production shift) in a way that is feasible (it can be achieved with the available resources) and very difficult to improve, i.e., it reaches one (or more) performance levels lying as close as possible to the optimal (i.e., not improvable) one. From a practical manufacturing production standpoint, the planning activity defines an overall plan that, taking into account the aggregate information about the production capabilities, specifies according to the company production policy the objectives for some suitable future timeframe. Such a plan represents a requirement (providing a set of constraints, establishing priorities, and so on) for the successive scheduling activity, which must define an operational plan, the schedule. This, in turn, specifies in detail how the production tasks should utilize the available resources.

Planning in manufacturing is usually considered an off-line activity; that is, it does not need to be executed in real time. Planning should not make use of detailed information about the current state of the production or resources because this increases the complexity of the decision-making process at this level and does not lead to any sensible improvement in the quality of plans. However, execution may include planning activities whenever these have a dynamic behavior affected by real-time events. For example, the supplier management support function as defined by McClellan [1] is considered as an MES support function whenever special management, such as outsourcing and just-in-time inventory management, is required.

Scheduling can be an off-line or on-line activity. This classification depends on the temporal characteristics of the scheduling problems, that is, on when the data needed to make the scheduling decisions are available, and on the incidence of unforeseen, stochastic changes that can occur for the entities involved in the scheduling situations. As a matter of fact, because a schedule must be a detailed operational plan, the lack of information or the continuous occurrence of changes can deprive an off-line plan of any relevance. In such cases, operational decisions must be taken on the fly — that is, as the things happen; to ensure that such decisions move in the right direction (i.e., that they generate schedules reasonably close to the optimal ones), they must be made on the basis of algorithms, rules, or policies whose performance has been evaluated (and possibly guaranteed) in an analytical or experimental (simulative) way. Planning and scheduling are thus two interrelated aspects of the same whole problem. Planning and scheduling mainly differ in the detail of the information they use to define a decision and the details included in the plans they generate.

Planning and scheduling have a hierarchical dependency: planning lies at an upper (planning) level and influences scheduling activities at a lower (operational) level, as results from the hierarchical organization of functions in the ANSI/ISA-95 or MESA-11 standards. At an even lower level lies the control activity; control is responsible for the execution of production on the manufacturing resources, particularly on the ones able to operate automatically. Moreover, control is the first source of feedback information about the course of production activities, providing the higher levels with the state of the single manufacturing operations and resources. Control is an on-line, real-time activity that can be considered to be strictly connected to scheduling, with a relationship similar to the one linking planning to scheduling. Planning first, and thereafter scheduling, determine the way in which production requests should use the available resources to yield high performance. On-line scheduling seeks to improve performance whenever *a priori* information on production requests is incomplete or unexpected events occur, and control ensures that the production requirements (e.g., the correct sequence of operations or the correct part program) are satisfied by the shop floor resources and monitors the actual state of production activities.

To exploit novel agent and multiagent technology in modern manufacturing systems effectively, the nature of the problems, or alternatively of the decisions, relevant to planning, scheduling, and control activities must be clearly understood. For this reason, three sections follow, each devoted to the analysis of a single activity, in which the problems are formalized and some references to classical (non agent-based) methods are recalled.

PLANNING, SCHEDULING, AND CONTROL IN MANUFACTURING FROM AN INTEGRATED PERSPECTIVE

This section faces the three fundamental problems, or phases, in MS corresponding to planning, scheduling, and control from an integrated perspective. As mentioned earlier, the three problems are strictly interconnected. On the one hand, planning and scheduling (P&S) differ for the aggregate vs. detailed way, respectively, in coping with production decisions and influence each other by imposing constraints (planning over scheduling) or by revealing inconsistency or actual infeasibility (scheduling over planning). Scheduling and control (S&C), on the other hand, may share the same temporal scope as the first one may be responsible for optimizing performance with on-line decisions; the second one, moreover, forces the shop floor devices to follow such decisions while respecting the correct production requirements. However, it should also be clear that planning, scheduling, and control deal with the "same" problem (obviously, from different, but contiguous, standpoints). Consider the block diagram of Figure 3.1 in which the basic manufacturing activities relevant to production management are reported in a three-layered hier-

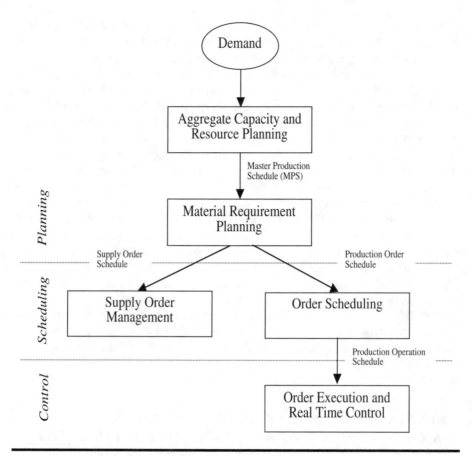

Figure 3.1   Main Activity Flow in Manufacturing Production

archy that evinces the three conventional planning, scheduling, and control decision levels.

In the figure, the typical flow of information, in particular related to decisions characterizing the production process in manufacturing, clearly highlights the strong link between the activities (blocks in the figure) devoted to planning and those devoted to scheduling. The lowest block in the diagram reflects the presence of the real-time control system, which is able to make things happen to the extent planned and to limit the effects of unforeseen events. P&S are closely related decision activities because they aim to determine feasible plans (schedules) that ensure high performance. S&C are connected in a real-time framework in which on-line scheduling decisions must be directly implemented. Real time is quite strict for S&C because revisions for schedules are only possible if unexpected events occur at the shop floor level and are revealed by the control system. Actually, the different time scope of planning, scheduling, and control activities is why, generally, a monolithic integrated planning, scheduling, and control system is not considered appropriate. The integrated perspective outlined here and used in the following sections of the chapter is then stronger in the case of P&S and weaker in that of S&C, because it can be established only in a specific timeframe.

Consider again the diagram in Figure 3.1. Even if it is a simplified view (disregarding, in particular, the supply and distribution aspects associated with production), the diagram represents the schema followed by most current information and decision support systems for managing production activities in manufacturing. Such a schema evolved from the first material requirement planning (MRP) systems developed in the 1970s and then was extended in subsequent manufacturing resource planning (MRP II) systems utilized until the mid 1990s; it forms the backbone of enterprise resource planning (ERP) systems and the currently "popular" supply chain management (SCM) systems.

These new generations of management information systems, even if exploiting innovations in information and communication technologies — especially the connectivity of systems through local or wide area networks — basically maintain the schema of Figure 3.1, using a hierarchical approach to tackle the complexity of the whole production management decision. For this reason, the adjectives *conventional* or *classic* are hereinafter used to identify such systems and approaches. The next three sections will strive to introduce the basic concepts needed to understand the problems and the classic approaches for planning, scheduling, and control in manufacturing. Clearly, these sections are not exhaustive in their coverage of the topics, for which a wealth of literature exists; they aim, rather, to provide the reader with the insights necessary for a greater appreciation of the appropriateness of an agent-based approach and the innovation that this may represent.

BASIC CONCEPTS IN PRODUCTION PLANNING

Planning in an MS is the process of determining tentative plans that specify what the MS should produce and what it should purchase in production periods in the near future. Note that this definition can be considered a simplified one because every activity in an MS that requires a plan for some future period, e.g., maintenance, or staff scheduling, needs planning as well. Here, however, the aspects relevant to production (and to some extent to inventory) management are stressed. The set of production periods over which a plan is defined is known as the planning horizon. During a planning process in an MS, future decisions related to the planning horizon must be taken. This decisional process results from an aggregation of decision variables, objectives, and constraints characterizing the decisions. Aggregation is a key term when considering planning: decisions are made by viewing time, production resources, and products as aggregate elements.

The Planning Decisions

The first purpose of planning is to establish the level of production for the MS products for the time periods composing the planning horizon. The detail of this decision is generally coarse because aggregate categories of products are considered. The definition of a plan for finished product and material inventories is also linked to such decisions, where the term material denotes any kind of raw material, component, or semifinished product required for the finished product. In this case, the levels of inventories must be determined for the different periods of the planning horizon and decisions must regard aggregate categories of the stored items. Resources different from materials used in the production, such as workforce or electrical energy, as well as machines and tools, are usually treated as fixed by the production planning process. This is particularly appropriate when the production resources are owned by the MS and are stable. However, some level of flexibility is possible, and plans can be defined that determine the production resource requirements, or simply their extra requirements, for the production periods.

The Information Needed by Planning

The stimulus to establish plans for finished products, materials, and, possibly, production facilities is the presence of a product demand for the corresponding time periods in the planning horizon. Similarly to what happens for the decisions about products and resources, demand is considered as aggregate. The problem of identifying reliable forecasts for product demand is clearly crucial because the planning decisions are driven by this information. In the context of *make-to-stock* production, demand is generated by a possibly stable forecast, because company stocks generally smooth market variations. By contrast, in *make-to-order* production, demand might be quite difficult to forecast effectively and the MS should be able to react promptly to the external stimuli represented by customer orders. These are two extreme situations; the reality may be a blend of both. In make-to-stock production, the unit value of products is usually not as high as in the make-to-order case; this feature highlights the risk of inconsistent planning especially in the potentially more costly situations. Several techniques exist to define the forecast of the demand: the more precise and detailed the required forecast must be, the higher the risk of mistakes is.

The classic methods to face planning span from deterministic and stochastic models for inventory planning to linear and mixed integer programming and to dynamic programming (a comprehensive review of which is found in Thomas and McClain [4]). These models assume that decisions depend on the forecast of demand for finished products and, apart from some mixed integer programming models, usually do not explicitly consider the induced demand for materials, components, and semifinished products. In addition, demand is generally averaged on the production periods and aggregated for product categories.

MRP was introduced to satisfy the need to take into account not only the exogenous independent demand, but also the dependent demand deriving from the independent one. This technique uses the notion of bill of materials (BOM) to disaggregate the finished product demand in the dependent demand for materials, thus making more detailed planning for production orders and supply orders possible. Figure 3.2 illustrates a simplified example of BOM for the production of a bicycle, with the MRP reports showing the induced production orders for one of the item's components. The figure shows the BOM tree hierarchy of the materials and semifinished products necessary for the manufacturing of a B01 bicycle, also reporting the relevant lead times. The two tables in the figure show, respectively, the result of the MRP computations for the finished product, the B01 bicycle, and the pipe components, the PP1 material; each table reports the net requirements and the corresponding planned production or purchase orders over a horizon of six production periods, computed from the gross requirements by taking into account available inventory and lead times. Among the numerous references on the topic, readers interested in further reading on MRP computations can refer to Nahmias [5].

MRP has been incorporated in manufacturing information systems since the 1970s. However, MRP does not perform any kind of optimization, but uses the available information relevant to disaggregated demand, stock levels, and, in particular, lead times to plan the quantity of finished and semifinished products to manufacture and the quantity of any other material needed by the production process that must be acquired from external suppliers for each production period. Thus, no optimization or capacity constraints are considered in MRP. The evolution of MRP systems to manufacturing resource planning (MRP II) systems was first characterized by the introduction of capacity requirement planning modules and by feedback toward the MRP module to guarantee reliability and actual applicability to MRP II plans (which, otherwise, could prescribe an unfeasible use of the resources).

One of the main difficulties in planning stems from the need to adapt plans to dynamic changes in demand; from another perspective, planning

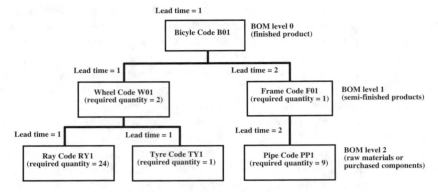

The BOM for a finished product code

Production Period		1	2	3	4	5	6
Gross Requirement			20				40
Scheduled Receipts				10			
Available Inventory	20	20	0	10	10	10	0
Net Requirement							30
Planned Order						30	

MPR computation for the finished product (Code B01)

Production Period		1	2	3	4	5	6
Gross Requirement				180			
Scheduled Receipts			40				
Available Inventory	80	80	120	0			
Net Requirement				60			
Planned Order		60					

MPR computation for a component (Code PP1)

Figure 3.2   Example of BOM and MRP Computations

in a case of stable demand can generally be considered a simple problem. One strategy to overcome such a difficulty lies in the continuous revision of the plan in order to make production (and associated purchasing) follow as closely as possible changes in demand. Taking into account that plans are an aggregate view of the production process (that need to be detailed to become an operational schedule) and that demand is generally affected by uncertainty, to follow changes in demand does not seem to offer an effective strategy. In addition, the policy of attempting to satisfy demand only on a period-by-period base could be neither feasible (e.g., for the lack of available resources in a given period) nor convenient (e.g., for the high cost of overtime in a given period).

Thus, the role of planning is to satisfy demand and to smooth its variations. This can be obtained by spreading production (and purchasing) over more than one production period and by setting inventory levels accordingly. The use of finished product and material stores can enable an MS to respond to peaks in demand occurring during a given period. The storage of materials can also extenuate the effect of unforeseen variations in suppliers' lead times that are the other important exogenous factors influencing planning reliability. The need for high interoperability with supply companies' information systems that B2B and E-procurement systems can offer also has the important effect of keeping suppliers' lead times under strict control.

The Planning Objectives

The strategy of satisfying demand variations by smoothing them entails costs that planning needs to minimize. The manufacture of different classes of products has a cost that can vary from period to period. Inventory also has a cost corresponding to the relevant investment and the required management. The possibility of using overtime to increase the production level clearly has a cost, as negligence of part of the demand not satisfied or backlogging. All of these costs must be appropriately estimated and, as is presumable, some of them are challenging. Planning seems to have many "enemies" ready to invalidate the effectiveness and reliability of its decisions. Planning parameters, such as time horizon, length of periods, levels of aggregation, replanning frequency, etc., must be fixed to adapt the planning method to the context at hand in order to grant validity to the planning prescriptions. On the other hand, part of the problem at times presents constraints that may be somewhat relaxed, thus reducing the complexity of the decision (e.g., production capability if overtime corresponds to a constraint that is violated by incurring a cost).

Some Classic Models for Production Planning

The literature offers numerous production planning problems and related solutions. The original paradigm in this context was the economic order quantity (EOQ) model [6, 7]. Based on the assumption that demand is constant and known and that a single item is involved, the EOQ model determines the dimension of the lots of items to be produced (or to be ordered from the suppliers) and the time instants at which production (or supply) orders must be placed, taking into account production (ordering) costs and inventory (so-called holding) costs. The main limitation of the EOQ model clearly results from its strong assumptions about the presence of a stationary and deterministic demand; of unlimited resources (so that a plan can always be fulfilled); and of a single item. Later models tried to take into account more realistic situations. The so-called Wagner–Whitin (WW) model [8] introduced dynamicity in the demand by subdividing the planning horizon into different production periods and considering different demand levels for each of them. The exact solution to the WW model is found by dynamic programming; nevertheless, even heuristic approaches (e.g., the Silver–Meal heuristic [9]) have been used when a model of this kind has been incorporated into MRP systems.

The introduction of capacity limitation usually corresponds to an increase in problem complexity, thus turning problems that can be solved by efficient algorithms into very difficult ones (for a more precise definition of the concept of complexity, refer to the section on the problem of computational complexity). Planning, however, works at an aggregate level so that capacity also usually measures the aggregate shop floor production capability. A number of models based on mixed integer programming (MIP) formulations have been studied to face production planning with limited capacity, multiproduct, multilevel, and multistage, as well as other characteristics. The relevant literature is extensive; the reader should consider Shapiro [10] as a basic reference, with other, more recent publications providing interesting resources for this class of models (such as Wolsey [11], Drexl and Kimms [12], and Clark [13]).

As an example, a well-known and studied MIP model is the so-called capacitated lot sizing problem (CLSP). In this problem, the levels of production for a set of products must be determined for a sequence of time periods in order to satisfy the external demand for the products in each period. The model takes into account (1) the production capacity of the manufacturing facility in each period; (2) the capacity requirement to manufacture the products; (3) the inventory cost for each product stored in a time period; and (4) the fixed cost required to set up the facility for manufacturing a product in a given time period. No efficient solution algorithms exist for the CLSP, which has been shown to belong to the class of NP-hard problems (again refer to the section on the problem of computational complexity for the definition of this class).

On the other hand, such a problem can be further complicated to mimic several realistic situations, e.g., multiple products, the possibility of backorders, the presence of different production facilities, and even the presence of a demand for raw materials, components, and semifinished products induced by the structure of the finished products, thus extending the model to incorporate the MRP functionalities. This last case is known as the multistage capacitated lot-sizing (MSCLS) problem and it also belongs to the class of NP-hard problems. The difficulty of efficiently finding the optimal solutions for nontrivial production planning problems is a challenge that researchers have tackled through several approaches, such as those based on evolution (e.g., Khouja [14], Kimms [15], and Disney et al. [16]); taboo search (e.g., Crauwels et al. [17] and Meyr [18]); or simulated annealing algorithms (e.g., Kim and Kim [19] and Barbarosoglu and Özdamar [20]). On the other hand, practitioners have usually preferred simple heuristics (e.g., Maes and Wassenhove [21] and Clark and Armentano [22]) as a way swiftly to determine feasible solutions that, hopefully, are not far from the optimal ones.

BASIC CONCEPTS IN PRODUCTION SCHEDULING

As already introduced earlier, scheduling corresponds to the activity of assigning production tasks to available resources and fixing the exact sequence and times at which each job must be performed. Thus, scheduling is a pivotal decision activity at the operational level because it specifies how the planned production must actually be carried out over time. This definition is clearly too vague to discuss properly the application of agent-based approaches to support scheduling in manufacturing and, in particular, to analyze their possible benefits and drawbacks. Consequently, this section gives a brief overview of the most relevant characteristics of scheduling problems in manufacturing in order to provide interested readers with the essential concepts and terminology.

A widespread community of researchers has studied scheduling problems over the last 50 years, and many efforts to find efficient approaches to tackle them will certainly be attempted in the future. The reasons for such a broad interest can be found in several aspects of this class of decision problems:

■ Scheduling problems have a high impact on actual production systems; efficient or smart solutions can drastically improve system performances, thus providing a valuable competitive edge.
■ The variety of different scheduling models that can be associated with production systems is quite extensive; no comprehensive model can be universally adopted and be worthwhile of investigation. Scheduling situations in general, and in manufacturing in particular, differ for the objectives that should be optimized and the constraints that must be satisfied. A telling example of this is found in the difference between the possible approaches to face scheduling problems in which all the relevant information is known *a priori* (the so-called off-line problems) and the ones for which a subset of such information becomes known on-line.

■ Apart from very simple cases, scheduling problems are difficult to solve. In fact, these problems generally belong to the class of combinatorial problems that cannot be solved by any efficient algorithm, i.e., the so-called class of NP-hard problems. For all practical purposes, this means that an optimal solution for most actual scheduling problems in manufacturing production cannot be found in a computation time that allows the exploitation of the solution. (As an example, a problem that must be solved in half an hour requires at least 2 days of computation to determine its optimal solution.) Instead of seeking out optimal solutions, approximate or suboptimal but still suitable ones should be identified. The interest of researchers in this field thus ranges from the study of problem complexity to the development of new exact, approximate, and heuristic approaches.

A comprehensive analysis of the scheduling problems arising in manufacturing production is clearly beyond the scope of this book. A selection of relevant, certainly incomplete, references can be found in the chapter references [23–26]. In the subsections that follow, basic terminology and key concepts about scheduling in manufacturing are introduced.

The Problem of Computational Complexity

Scheduling in manufacturing deals with the efficient use of resources to perform the various activities that make up a production process and is often referred to as finite capacity scheduling because the resources involved have limited availability. Scheduling corresponds to operational decisions about the way to perform routine or day-by-day processes efficiently in order to render manufacturing production economically convenient and to ensure fulfillment of customers' requirements. Despite their regularity and the fact that they do not have an impact upon company strategy or mid-term planning, operational scheduling problems in most cases present dramatic challenges that discourage their solution by means of exact approaches in real application contexts. Scheduling problems are, in fact, combinatorial problems, i.e., problems for which the number of solutions is finite and enumerable, yet for which an efficient solution is very often unavailable due to their computational complexity.

Since the 1970s, the theory of computational complexity has provided an essential tool to evaluate *a priori* the possibility of designing algorithms that require acceptable computation times to solve a problem optimally. An outstanding reference text on the theory of the computational complexity for computer scientists and operational researchers is *Computers and Intractability* [27]. Hereinafter, only a quite informal definition of the concepts of easy, or well-solved, problems and difficult, or intractable, problems is introduced.

Computational complexity theory provides a methodology to evaluate the difficulty of solving problems in terms of the amount of time and memory needed to achieve this end. Focusing on the computation time, the theory quantifies the problem's difficulty independently from the specific computer hardware used to run the algorithms because the time needed to solve a problem is measured by the number of elementary operations required (e.g., additions, reading, writing, shift, and so on). Such a measure is provided by a so-called time complexity function of the input of a problem (i.e., the size and length of the data needed to describe a generic instance); in particular, only the highest order of such a function is considered because this is representative of the worst run of the algorithm. The difficulty of a problem is then associated with the order of the time complexity function of the best algorithm among the ones currently available to solve any instance of such a problem.

Easy or well solved are those whose time complexity function has a polynomial order, usually denoted with $O(p(k))$, where $p(k)$ is a polynomial function of k, the length of the input of a generic instance of the problem. Such a problem belongs to the class P that includes the problems for which an efficient, i.e., polynomial time, algorithm is available. It must be noted that only polynomial time algorithms with a small order are practically useful because orders greater than $O(k^4)$ could be unacceptable to solve large instances of a problem optimally. Unfortunately, only a few combinatorial problems belong to the class P (or, as is said, they are in P).

Another interesting class of problems is the NP class. This class is made up of problems whose solution, once provided or guessed, can be verified in a polynomial time. This class is larger than P and, in particular, it is assumed that $P \subseteq NP$. NP stands for *nondeterministic polynomial* because the problem in NP can be "ideally" solved by a fictitious nondeterministic polynomial algorithm that in some way guesses the solution and proves its correctness in a polynomial time. A problem π is said to be NP-complete (NP-C) if $\pi \in NP$ and any other problems $\pi' \in NP$ can be reduced to π by means of a polynomial transformation (i.e., not changing the order of the problem complexity). In practice, this means that an algorithm that is able to solve the instances of the problem π can be used as a subroutine in an algorithm that, with a complexity of the same order, is in turn able to solve the instances of any other problem π' in NP. For this reason the NP-C problems are considered as difficult as any other problems in NP.

Finding a polynomial time algorithm to solve one NP-C problem means discovering a way to solve any problem in NP efficiently — in other words, proving that $P = NP$. Unfortunately, even if it has never been demonstrated, this possibility is assumed to be extremely unlike. Thus, the only types of algorithms that may be used to solve NP-C problems optimally have a time complexity function of an exponential order (as, for example, $O(2^k)$), such that NP-C are computationally intractable due to the extremely rapid rise of the computation time for instances of realistic dimensions. A technical point to bear in mind is that the theory of computational comlexity regards problems in decision form (i.e., problems for which the true answer between "yes" or "no" must be determined) rather than problems in optimization form (i.e., optimization problems); however, it is easy to devise an algorithm to solve an optimization problem that invoke a polynomial number of times as a procedure an algorithm solving the correspondent problem in decision form.

An optimization problem, like a scheduling problem, to which an NP-C problem can be reduced by a polynomial transformation, is said to be NP-hard, meaning that it is at least as difficult as the NP-C problem. Thus, saying that most scheduling problems are NP-hard means that, apart from instances of a very small size, those problems cannot be optimally solved by any efficient algorithm or, equivalently, in an acceptable computation time. Note that the theory of computational complexity deepens the characteristics of combinatorial problems, providing the definition of many other classes that have not been mentioned here (a valid reference is Garey and Johnson [27]). The purpose of the short and informal introduction of this section aims solely to highlight why practical scheduling problems are very often tackled only by means of approximate or heuristic algorithms; such algorithms, in fact, provide in a polynomial time an acceptable suboptimal solution; in this connection, multiagent applications represent a novel approach.

Basic Terminology and Concepts for Manufacturing Scheduling

In this section, the main terminology and concepts relevant to scheduling in manufacturing contexts will be reviewed. However, for a deeper analysis of this topic, the interested reader can refer to the extensive literature available, among which some valuable samples are Blazewicz et al. [26], Pinedo [28], and Brucker [29].

Many different scheduling problems arise in different manufacturing industries. The point that must be stressed is that real scheduling problems cannot usually be faced without simplifying some of their specific aspects. As an example, the amount of time needed to perform an operation, the so-called *processing time*, is often considered a constant even if, in most cases, this time span can be slightly modified by stochastic events. However, modeling the possible variations in processing times often increases the difficulties of the scheduling problems without sensibly improving the quality of the solutions.

Finite capacity scheduling problems basically require that a set of job J is executed by a set of machines M in a feasible way, i.e., satisfying a given set of constraints and possibly optimizing certain performance criteria. Different scheduling problems are thus characterized by the specific objectives and constraints that should be considered. Scheduling problems are first defined next by focusing on the most relevant constraints affecting the feasibility of the solutions; thereafter, performance measures and solution approaches are introduced.

Feasibility-Based Characterization of Scheduling Problems

The jobs in a scheduling problem usually correspond to production orders, and their execution can require the processing of single or multiple operations or tasks. In the case in which the single jobs are not constrained by precedence relations among them, they are said to be *independent*. In the case of multioperation jobs, a generic job j can be divided into n_j tasks, denoted with T_{1j}, T_{2j}, ..., $T_{n_j, j}$, which may require different kinds of processing and among which a set of precedence relationships may exist. The term *task* is thus used to indicate a single operation that is needed at a certain stage of a production process and may be preceded or followed by other operations. A job indicates a production activity as a whole and can be associated to production orders or lots. In the literature, three main scheduling models are usually considered:

- In the *flow shop* model, every job is composed of the same number, n, of tasks; here, the tasks of any job should be processed in the same order, each on a different machine, thus creating flows of jobs among the machines, i.e., T_{hj} must precede $T_{h+1,j}$, for $h = 1, ..., n - 1$; as a result, the number of machines is assumed to be equal to n.
- The *open shop* model, similarly to the flow shop model, considers the same number, n, of tasks for each job and n different machines to provide processing; in this case, however, the processing order of the tasks on the machines is not fixed.
- In the *job shop* model, the number, n_j, of tasks making up a job j is arbitrary, as is the structure of the precedence relationships among the tasks of different jobs.

The set M of machines represents the fundamental production resources whose characteristics range from very simple to very complex ones. In particular, two basic configurations for the machines can be reviewed: *parallel machines*, sometimes called processors (i.e., machines that can provide the same kind of processing and are able simultaneously to serve one single task each) and *dedicated machines* (i.e., machines that can provide a specific kind of processing).

Frequently, in the context of manufacturing the processing of a task cannot be interrupted and then resumed later by the same or other machines; in other words, so-called task *preemption* is not allowed. Both these basic configurations are characterized by the speed of the machines in processing the tasks. If the parallel machines have the same speed, s, they are called *identical*; otherwise, if the machines have a different speed, s_i, $i \in M$, which is independent of the particular task processed, they are called *uniform*. If their speed, s_{ij}, also depends on the job $j \in T$ (or task) processed, they are said to be *unrelated*. Dedicated machines can also be associated with complex production cells, thus forming a so-called flexible manufacturing system (FMS). In this case, a machine actually corresponds to a set of resources, including, for example, different tools; after an appropriate set-up, this allows the machine to execute different kinds of operations. Therefore, scheduling models for an FMS should take into account the set-up time that must be spent to switch between different processing operations in order to identify a feasible solution.

The main information that can be associated with a generic job (or task) $j \in J$ can be summarized as follows. The processing time, p_{ij}, is the time needed to complete the execution of job j on the machine $i \in M$; if the machines are identical, then $p_{ij} = p_j$ for every $i \in M$, where p_j represents the standard processing requirement of the job. If the machines are uniform, then $p_{ij} = p_j/s_i$ and, if they are unrelated, then $p_{ij} = p_j/s_{ij}$. The ready time, r_j, represents the time instant at which the job is ready to start its execution, e.g., due to the availability of raw material. The due date, d_j, indicates the time instant at which the job should be completed without incurring penalties; such a date is usually agreed upon with the customer. The deadline, dl_j, represents the time instant at which the job must be completed (if the job deadline is exceeded, the customer rejects the corresponding order). The priority or cost of the job, w_j, is used to weight the job's features within the performance measures to be optimized.

A set of precedence relationships among the tasks composing a job constrains its production process. In general, this set can be represented by a directed acyclic graph, $G_p = (V,A)$, where the set of vertices, V, is associated with the tasks, and by a directed arc, $(b,k) \in A$, if the task b must be completed before starting the processing of the task k. A set of additional resources may be needed to process the tasks and may include resources that can be allocated to the tasks in a continuous way, such as electric power, and discrete resources such as tools. Scheduling problems characterized by the presence of additional resources are identified as resource constrained.

A schedule **S** is an assignment of the tasks to the m machines; this partitions the set of tasks or jobs into m subsets S_i, $i = 1, \ldots, m$, respectively, including the tasks assigned to the machine $i = 1, \ldots, m$, i.e.,

$$S = \bigcup_{i=1}^{m} S_i \, .$$

In addition, a schedule specifies the sequence of the tasks on the machines; thus every set S_i is completely ordered, i.e., $S_i = (T_{i1},\dots,T_{ib},T_{i,b+1},\dots,T_{i,n_i})$, where task T_{ib} is sequenced on machine i before task $T_{i,b+1}$, $b = 1, \dots, n_i - 1$, where n_i is the number of tasks sequenced on machine i. Finally, a schedule determines a timetabling, i.e., the starting time, S_j, for every task $j \in J$, in particular such that $S_{b+1} \geq S_b + p_{ib}$, for each pair of tasks T_{ib}, $T_{i,b+1}$ consecutively sequenced on the same machine i, where p_{ib} is the processing time of T_{ib}. As a consequence, the following quantities are determined:

- C_j, the completion time of the task j, corresponding to $C_j = S_j + p_j$
- F_j, the flow time of the task j, corresponding to $F_j = C_j - r_j$
- L_j, the lateness of the task j, corresponding to $F_j = C_j - d_j$
- E_j, the earliness of the task j, computed as $E_j = $ max $[0, \, d_j - S_j - p_j]$ = max $[0, \, d_j - C_j]$
- T_j, the tardiness of the task j, computed as $T_j = $ max $[0, \, S_j + p_j - d_j]$ = max $[0, \, C_j - d_j]$

A schedule is feasible if it defines an assignment, a sequencing, and a timetabling that satisfy the problem constraints. In the case of nonpreemptive scheduling, this corresponds to processing every task j on a single machine without interruption in a time window $[r_j, \, dl_j]$ and without violating any precedence relationship, so that every machine processes only a single task at a time and any other possible condition on additional resources is respected. A schedule is usually represented by a Gantt chart, which is a two-dimensional diagram reporting time on the abscissa and the machines on the ordinate, and showing the time intervals in which the tasks are processed by means of boxes corresponding to the assigned machines.

As mentioned earlier, an important feature of scheduling problems regards the *a priori* availability of information about jobs. Problems are deterministic if the information about jobs is available *a priori*. When part of this information becomes available only during the execution of jobs, an on-line scheduling problem must be faced. Different classes of on-line scheduling have been defined in the literature, depending on which information is revealed on-line and how this comes about. For example, the scheduling one-by-one model assumes that the information about a new job becomes available just after the previously considered job has been scheduled. On the other hand, the arrival-over-time model assumes (as its name implies) that jobs can arrive and reveal their characteristics over time, but that their scheduling can also be delayed. An exhaustive survey of on-line scheduling models and algorithms is provided in Sgall [30].

Dynamic scheduling models occur whenever jobs arrive over time, but some statistical information about their arrival is known *a priori* (e.g., a probability distribution of arrival for different kinds of jobs is given). Dynamic scheduling problems can be faced by means of so-called dispatching rules, whose performance can be analytically studied in a few simple cases through queuing theory results or, most of the time, by means of a simulation campaign. Usually, in practice, solutions identified by means of off-line algorithms or through dynamic dispatching rules need to be modified on-line because things do not often happen as expected. Thus, depending on the considered situation, predefined solutions or policies for scheduling problems can be alternatively implemented, slightly modified, or almost completely ignored in a real-time operational context.

Performance-Based Characterization of Scheduling Problems

The performance of a schedule is evaluated by computing the value of an objective function, and the nature of such a function is an important classification key for scheduling problems. The effectiveness of a scheduling algorithm depends on its ability to explore the scheduling solution space and find a feasible, not far from optimal, solution in an acceptable timeframe. Most of the literature about scheduling deals with single-objective problems, but in some situations more than a single objective should be considered because a multiobjective scheduling approach is needed. A schedule is deemed optimal if it provides a value for a specified objective function that cannot be improved any further.

Several different performance criteria are used to evaluate the quality of a schedule. Specifically, *regular* and *nonregular* objectives are distinguished in the literature. If $Z(C_j)$ is a generic schedule objective to be minimized, expressed as a function of the completion times of the jobs, then Z is said to be regular if $Z(C_j) \leq Z(C'_j)$, where $C'_j \geq C_j$ for every $j \in J$. This means that, given an assignment and a sequencing of the jobs on the machines, a regular objective cannot be improved by delaying the completion (or the start) time of the jobs. Objectives (to be minimized) with such a characteristic are, for example:

■ The *makespan* of the schedule, i.e., its length,

$$C_{\max} = \max_{j \in J} C_j$$

■ The mean (weighted) completion time,

$$\overline{C}_w = \frac{\sum\limits_{j \in J} w_j C_j}{\sum\limits_{j \in J} w_j}$$

■ The mean (weighted) flow time,

$$\overline{F}_w = \frac{\sum\limits_{j \in J} w_j F_j}{\sum\limits_{j \in J} w_j}$$

■ The maximum tardiness,

$$T_{max} = \max_{j \in J} \ T_j$$

■ The mean (weighted) tardiness,

$$\overline{T}_w = \frac{\sum\limits_{j \in J} w_j T_j}{\sum\limits_{j \in J} w_j}$$

■ The number of tardy jobs, i.e., jobs completed after their due date

The optimal schedule for one regular objective can be found by limiting the search space to the so-called semiactive schedules, i.e., schedules for which, given an assignment and a sequencing of the jobs on the machines, no job can be started earlier without modifying the assignment and sequencing or violating any constraints on precedence conditions or ready times. Recently, the nonregular objectives have become the subject of increasing interest. These objectives can be improved differently from the regular ones — even by delaying the completion of jobs — i.e., by inserting idle times in machines' schedules.

An interesting case of a nonregular objective to be minimized corresponds to the total (weighted) earliness and tardiness, with earliness and tardiness weights, we_j and wt_j, $j \in J$, $ET = \Sigma_{j=1,...,n}(we_j \cdot E_j + wt_j \cdot T_j)$. The nonregularity of such an objective depends on the penalty for the earliness of the jobs, that is, on the wish to minimize the total weighted deviation of the completion times from the due dates. The purpose is thus that of reducing not only the penalties for the late delivery of an order with respect to the due date, but also the costs pertaining to an early production that correspond to early expenses for supplies and missed (delayed) profits for storing raw materials and finished goods. Although the adoption of the just-in-time (JIT) strategy has raised the popularity of earliness and tardiness (E/T) minimization, research on E/T schedules still seems immature compared to the results achieved with efforts on regular objectives. Important surveys on E/T models and scheduling with inserted idle times are retrievable in Baker and Scudder [31] and Kanet and Sridharan [32].

Some Remarks about Classic Solution Approaches

The main difficulty with scheduling problems, as already pointed out, is that, apart from very simple cases, they belong to the class of the NP-hard optimization problems. This fact has made the use of exact solution approaches (such as the ones based on mixed integer programming, branch and bound or dynamic programming) unsuitable for the myriad of problems involving a number of jobs and machines of practical interest. Approximated and heuristic approaches have been designed and analyzed in order to devise algorithms providing suboptimal solutions as close as possible to the optimal ones in acceptable computational time.

The literature about scheduling is so extensive that we can only recommend the texts on the topic already mentioned in this section. Here, we would cite only a widely used family of approximated approaches that, for their simplicity, are often used in practice: the list schedule algorithms, which generate a solution using a generally simple priority rule. For example, very popular rules are the longest processing time (LPT); shortest processing time (SPT); earliest due date (EDD); and first in, first out (FIFO), which order jobs according to one specific parameter (the processing time, due date, or ready time) and assign and sequence jobs in that order on the first machine available to process them. Many studies have been conducted to analyze the performance of such rules (see, for example, Panwalkar and Iskander [33] and Ferrel et al. [34]). In particular, the rules used in list schedulers produce solutions that are α-approximations of the optimal one; that is, they provide an objective value α times worse than the optimal one, with $\alpha \geq 1$.

The interest in such simple scheduling approaches is also justified by the fact that the rules can often be used in an on-line or dynamic context. One of the famous algorithms used to tackle the NP-hard problem of scheduling a set of independent jobs on a set of m parallel identical machines, with the objective of minimizing the makespan, is the Graham list scheduling algorithm [35]. This rule has been proved to be a $(2 - 1/m)$-approximation and is particularly interesting in the setting of on-line scheduling because it is considered the first proof of the so-called competitiveness for an on-line scheduling rule. As a matter of fact, the concepts of competitiveness and competitive analysis have been introduced to evaluate the performance of an algorithm operating on-line by means of a sort of worst-case analysis: an on-line algorithm, A, is said to be α-competitive for a problem, P, if, for each instance i of P, the cost provided, $C(A)$, is bounded so that $C(A) \leq \alpha \cdot C(\text{Opt}) + \beta$, where $\alpha \geq 1$ is a positive constant named *competitive ratio*, Opt denotes the optimal off-line algorithm for P, and β is an appropriately fixed constant offset. Thus, when the sequence of independent jobs to be scheduled on m parallel identical machines is not known *a priori*, the on-line use of the Graham list rule provides a $(2 - 1/m)$-competitive algorithm.

The competitive analysis is therefore an analytic way of measuring the performance of an on-line algorithm (for a reference see Borodin and El-Yaniv [36] and Irani and Karlin [37]), but presents two drawbacks: it may not be simple to perform (an on-line algorithm may not be competitive, i.e., the worst case it produces may not be bounded from above), and it may yield an index quite far from an average real one. For this reason, it may often be appropriate to evaluate the competitiveness of an algorithm statistically by averaging the competitive ratio on a number of randomly generated on-line instances, similarly to what is done to evaluate the competitiveness of randomized on-line algorithms, in which the ratio is averaged on the random choices made by an algorithm for each problem instance.

BASIC CONCEPTS IN PRODUCTION CONTROL

Although it is a general term, *production control* is more often associated with the lowest level functionalities of a manufacturing system, such as shop floor processes. Shop floors in modern manufacturing enterprises are already endowed with versatile production means (e.g., numerically controlled machines, automatically guided vehicles, and automated warehouse), thus allowing implementation of the many characteristics required for modern agile manufacturing: dynamic reconfiguration, improved productivity, and high operational flexibility [38]. However, by definition, the plant system is for the most part stable in configuration, and new approaches are required that can meet the challenge of rapid, adaptive response [39].

In a shop floor system, a set of distributed or digital control systems (DCSs) is present in order to control and monitor the process. In general, most manufacturing processes consist of one or more of the following process types: discrete parts manufacturing; assembly; batch processing; or continuous processing. The process type included in the overall manufacturing process mainly depends on the products produced. To a greater extent, discrete parts manufacturing dominates manufacturing activities in which integer numbers describe the amounts of products, such as in aircraft, automotive, electrical, and electronic manufacturing. Assembly is usually related to sequential process activities (such as in the flow shop scheduling model) and most often dominates aircraft, motor vehicle, and computer manufacturing. Batch processing is generally related to products following a production recipe and is usually applied to food, beverage, and pharmaceutical manufacturing. Continuous processing is more generally prevalent in manufacturing activities in which real positive numbers describe the amounts of products that are normally related to a continuous mass, for example, as usually happens in chemical and petroleum manufacturing. Next, these manufacturing processes are briefly introduced, and some key elements of their activities are detailed. Note, however, that because two or more of these processes may coexist in manufacturing production, these elements should be viewed as characterizing — and not limited to — each process.

The Manufacture of Discrete Parts and Assembly Processes

The process of manufacturing discrete parts typically involves multiple machines. These machines may perform altogether different activities as part of a quite complex set of tasks, for example, raw materials movement and storage; finished parts fabrication; packaging; storage; shipment; etc. A human–machine interface (HMI) — also known as man–machine interface (MMI) — generally acts as a link between the human operator and the machine, usually in the form of a device that displays machine or process information and provides the means for data entry or commands. An HMI often takes the form of a software product with graphic interfaces equipped with visual and audio alarms.

Although the machines used may vary widely, all of them need some type of control system. A relatively common key element of a control system is the programmable logic controller (PLC). A PLC is an industrial computer able to dialogue in hard real time with a certain number of machines through its input–output (I/O) system. The I/O system allows a PLC to receive inputs from switches and sensors related to signals representative of the actual condition of one or more machines and to generate outputs to actuating devices in order to control the operations of one or more machines. In addition, a PLC may also communicate with other devices (such as operator interfaces, variable speed drives, computers, and other PLCs) via one or more communication ports. A PLC control system can also be coupled with computer numerical control (CNC) when the manufacturing process is required to perform complex and exacting operations on machines such as lathes, grinders, laser cutters, etc.

A traditional PLC works by continually scanning a program through a sequence of three important steps. First, a PLC determines the states of each input device and records them in its memory. Next, the PLC executes the program one instruction at a time. Finally, the PLC updates the status of the outputs according to the inputs of the first step and the results of the second step. After the third step, the PLC goes back to step one and repeats the steps continuously. The duration of a scan time (defined by these three steps) depends on the processor speed and the length of the program. More recent PLCs are able to combine the discrete and analogue I/O control found in "hard PLC's" with the powerful data handling, programming, networking, and open architecture features of computers.

One typical process often coexisting with discrete parts manufacturing is the assembly process. Manufacturing assembly processes generally refer to a sequence of mounting operations through a series of assembly stations. Units being assembled are moved among stations via some transport mechanism, e.g., conveyor belts, trolleys, AGVs, robots, etc. Any specific assembly station may include a mix of manual and machine assembly operations. Many aspects of assembly processes are similar to discrete parts manufacturing and, in fact, many production processes combine them. Also in this case, PLCs are key elements of the control process, which is in general mainly devoted to data collection from switches and sensors in order to switch motors on or off and/or to control their velocity to coordinate the steps of the assembly process.

However, the geographic spread of an assembly process on the shop floor presents two main characteristic needs. The first entails the need for additional HMIs more evolved than traditional HMIs, that respond, for example, to the need to resemble information graphically and in databases related to alarms and/or processed data coming from PLCs and other DCS. These kinds of HMIs are so evolved that they can be used in the traditional operation systems of conventional workstations such as PCs. Supervisory control and data acquisition (SCADA) systems provide an example of the last stage of evolution of such utilities. This acronym generally refers to a category of process control software applications that gather data in real time from remote locations in order to control equipment and conditions.

The second, critical need has to do with the assurance of a rapid information flow from the several locations of the assembly process. Different devices such as PLCs, drives, computers, and operator interface systems must be interconnected by one or more local area networks (LANs). Although the LAN concept in manufacturing is the same as that of LANs found in more traditional environments such as offices, LAN used in industrial applications must be able to operate reliably under more stressing conditions, for example, those arising from a high level of electrical noise and a greater range of temperature and humidity that may be found in the plant. Industrial LAN specifications (and related standards) differ according to the requirements of the application, which comprise, for example, the data rate; number of devices to be connected; required reliability; compatibility with other networks; costs; etc. It is generally not possible for one network standard to maximize all these features, so a layered communication architecture is often used at the shop floor level, where several, possibly open, standards work together in different situations. Figure 3.3 shows an example of a simplified LAN architecture at the shop floor level.

A LAN in a manufacturing setting can be divided into three main layers that differ for bandwidth and target of application: field/process/device; control; and management. The lowest of these layers, field/process/device, is often referred to as a fieldbus network. Generally speaking, it consists of a digital network designed to replace traditional analog signals and

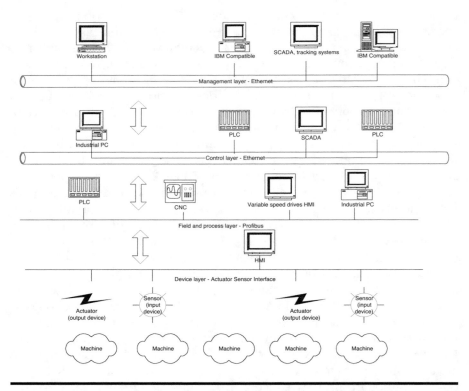

Figure 3.3 Example of Layered LAN Architecture in a Manufacturing Context

represents the link among field devices such as transducers, sensors, actuators, or controllers. Standard examples are the actuator sensor interface (ASI), Profibus DP for device networks, and Profibus FMS and IEC Fieldbus for the field/process layer.

The control layer generally interconnects equipment controllers such as PLCs and industrial PCs and fulfills the key task of providing a reliable, high-speed, and consistent exchange of time-sensitive control signal information among a moderate number of specialized computing devices over a relatively large area. The inclination here is to expedite the integration of this layer with the management layer so that similar network architectures are stressed. The so-called Ethernet family protocol (IEEE802.3 and IEEE802.11 in case of need of wireless connections) is thus evermore present as a data link and physical protocol, and TCP/IP is increasingly used to allow typical intranet/Internet connections and applications.

Finally, the management layer — the highest level of communication in the plant — is based on traditional LAN architectures. MES works in the plant using this network layer and can dialogue with shop floor machines through the control layer and with ERP and other IT systems through a traditional wide area network (WAN), which enables the connection of all the corporate units and plants.

Batch and Continuous Processes

Industrial batch processes are relatively straightforward because their control procedure, albeit scaled up to produce a larger quantity of material, follows a recipe similar to the process of baking a cake. Foods, drinks, medical products, paints, and building materials are examples of products manufactured using batch processes. Tracking a batch process in order to assess the genealogy of the product — that is, the record of the state of the product throughout all the important phases of the recipe — is a typical requirement of this production. ANSI/ISA S88 is a well-perceived standard providing suggestions and examples, but not a direct list of requirements, on three main defined elements of batch processes: instructions on how to make a product (recipe); the physical tools required in production (equipment entities); and the methods to link a recipe to equipment entities (control activities).

Continuous processes share some similarities with batch processes and both often coexist. In continuous processes, more emphasis is put on precision: ingredients that must be combined in precise ways at precise points in the process. Precise process control must be maintained to ensure product quality and operational safety. The need for precise process control is a must for some industries (namely, chemical, petrochemical, and metallurgic) that make extensive use of continuous processes. Many other industries, on the other hand, use continuous processes in some operations that are not directly linked to the production process (e.g., treatment of residue waste from the production process).

One important difference with respect to discrete parts manufacturing is that batch and continuous processes use mainly analogic data, which vary with continuity within a specified range and are usually representative of the physical variables of the process, such as temperature, pressure, rate of flow, weight, or any other important feature. Batch and continuous processes require continuous monitoring. Corrective control actions are often required to ensure that the process stays within specifications, for example, by measuring a value, comparing the measured value to a desired value or set point, and correcting for the error in a so-called closed-loop control.

Depending on the complexity of the process being controlled, several approaches can be used for process control. A small batch process can be controlled by just one PLC. A representation of the process showing its current status and a history of data and alarms recorded at various points in the process is often provided by a SCADA system networked to the PLCs.

Integration of Control Processes

Control processes are generally considered the terminal leaves of a highly centralized information system. In a simplified view, sensors/actuators are usually connected to some PLCs, which must be connected to SCADA and MES systems, and all these systems must be connected to the ERP system. This view has some limitations with regard to agility because supply chain management influences the dynamics of production through the ERP system; that is, production can be affected only after the ERP negotiation. As introduced in the previous chapter, it is plausible that, in the near future, this layered centralized architecture will collapse into one monolithic distributed architecture (it does not matter whether an ERP or an MES or a SCADA or any other software system will enclose the others) and that this new software will be able to select and to negotiate agreements with other companies in order to exploit agility through the use of a clustered network of companies.

In the meantime, one priority of MS is to introduce agility in the current classic layered architectures; at the moment, the lack of standards for the integration of business tools with shop floor systems presents the biggest challenge to overcome. The Instrumentation, System and Automation Society (ISA) recently introduced the ANSI/ISA-95 Enterprise-Control System Integration standard to "regulate" this field (Figure 3.4). The ANSI/ISA-95 [3] standard initially defines a common set of terms and definitions of the

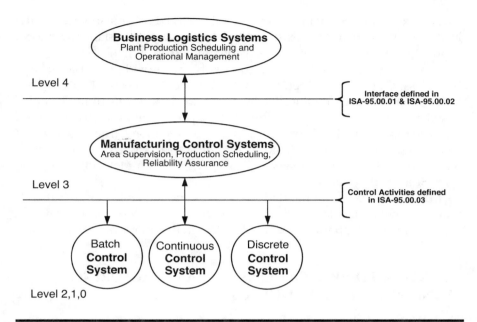

Figure 3.4   ISA 95 Functional Levels (From ANSI/ISA-95.00.01-2000, Enterprise-control system integration, part 1. models and terminology.)

information and activities associated with logistics and manufacturing integration. The terms include definitions of the activities of business logistics systems, activities of manufacturing control and coordination systems in multiple levels of detail, and the information that must be exchanged between these activities. Thereafter, ANSI/ISA-95.00.02, defines in more detail the attributes of the object models included in Part 1. The aims are to reduce the need for customized integration solutions; to simplify multivendor integration; and to improve the reusability and transportability of functions across the enterprise. The availability of information everywhere and whenever in an electronic form is the reason why the data processing systems of different levels must be integrated. ANSI/ISA-95.00.03 (not available yet) should dedicate greater attention to control activities.

Another road toward integration follows technological aspects. In fact, ERP and MES need accurate, real-time data from the shop floor. The ERP/MES decisions are sent back to the shop floor and to the related control system in order to accomplish production. A technological solution such as the distributed object-based approach can provide standard interfaces between different level applications; many research efforts are devoted to this concept. One main goal of these standard interfaces is to resolve the integration problems by enabling communication among different software packages without the need for customized codes or drivers.

For example, the main objective of the IEC 61131-3 standard [40, 41] has been to standardize existing PLC languages. The standard allows proprietary function blocks to be programmed in non IEC 61131-3 languages such as C++ so that it is also possible to provide extensions fairly "seamlessly," e.g., packaged as function blocks such as IEC 61131-7, Fuzzy Control Programming. Examples of integrating distributed object-based technologies that are increasingly applied in production control are: Microsoft COM/DCOM; OPC (OPC Foundation) using OLE/COM technology of Microsoft; CORBA by the Object Management Group (OMG); and Java technologies such as Java Remote Method Invocation from Sun Microsystems.

In conclusion, bear in mind that standards and technologies are valuable tools for the integration of control and business processes. However, a general architecture framework enabling integration must make use of them.

AGENT-BASED APPLICATIONS IN MANUFACTURING PLANNING AND SCHEDULING

The purpose of this section is to summarize the current trends of applications of MASs to P&S in manufacturing. As previously explained, P&S activities are two contiguous aspects that are often confronted in an integrated way by MAS applications. The objective is to examine the possibility of exploiting the different modeling alternatives, in terms of system architectures; agent roles and types; and interaction protocols, in order to define P&S systems for manufacturing. This section reviews some key concepts arising from successful approaches proposed in the literature over the last 10 years. A more complete and detailed summary is provided in Chapter 6.

The basic architecture schemes that have been introduced and adopted for MASs in manufacturing correspond to the three main agent organizations that follow [42, 43]:

■ *Hierarchical architecture*: agents operate at the different levels of a hierarchy that basically reproduces the hierarchical distribution of responsibilities and control in a manufacturing organization. Therefore, higher level agents control the activities of lower level ones, and communication usually takes place through the vertical levels of the hierarchy. This kind of model does not exploit the potentiality of an MAS, but adapts agents to a pre-existing "bureaucratic" organization. Flexibility is thus clearly reduced and, rather than a decentralized and distributed solution typical of agent-based systems, it takes the form of a monolithic centralized system.

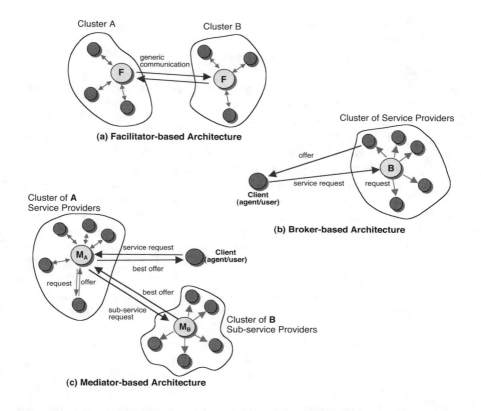

Figure 3.5   MAS Federation Architectures Based on (a) Facilitators; (b) Brokers; and (c) Mediators

- *Federation architecture*: agents are organized in small communities or clusters, depending on their functions, and other agents play the role of coordinators of the activities in several ways. In particular, alternative federation architectures can be based on the presence of facilitator, broker, or mediator agents, as depicted in Figure 3.5. Facilitator agents are used to coordinate communication between agents in the MAS, obviating the overwhelming increase of message exchanges if any agent is allowed to broadcast messages directly to the whole agent community. Agents are subdivided into clusters according to some shared property (e.g., they could provide a common service or they could be associated with a common class of physical entities like shop floor resources such as AGV); the facilitators act as a sort of communication gateway to rationalize the message traffic in the MAS. The facilitator allows the direct communication only of agents belonging to the same given cluster, and communications among different clusters are implemented by means of an exchange of messages through their facilitators.

Broker and mediator agents in federation architectures play more active roles, which extend the communication function of facilitators. Similarly to facilitators, broker agents are associated with clusters of agents, but this time specifically on the basis of the kind of service provided by the agents. Brokers thus accept requests for the specific kind of service and act as intermediaries between the clients and the service providers, ideally creating a communication link among them. They are also responsible for monitoring the presence of agents in the system, thereby providing a specific kind of service. In mediator-based architectures, mediator agents manage the requests for services similarly to broker agents; however, they also coordinate the activity of server agents and select from among the possible alternatives those server agent configurations that can more efficiently serve a request. Such a scheme can be particularly suitable to situations in which a service is provided by the coordinated actions of a set of heterogeneous agents: whenever a complex service request requires more service provider agents, a mediator distributes the requests to the appropriate set of providers directly or through some lower level mediators; collects the offers from the alternative providers (or relevant mediators); and selects the best one, for example, on the basis of a performance index like the service cost.

It should be borne in mind that these alternative schemes for federation architecture represent only reference points: other possibilities can be developed that involve classes of "social" agents like facilitators, brokers, or mediators with intermediate or even extended functions. However, in a federated MAS architecture, control is distributed over the enterprise network and, referring to the different types of distributed decision control architectures, it may be modeled as a so-called heterarchy. In such architectures, agents communicate as peers; because there are no fixed master–slave architectures, each type of agent can be replicated many times and global information is eliminated [44].

■ *Autonomous agent architecture*: agents are responsible for specific functions or services; they are not assigned to any particular agent organization nor are special coordinator agents introduced. This distributed decision architecture is certainly the one with the higher level of flexibility because any time a new entity providing a service in the MS becomes available, a new autonomous agent managing that service is activated. Conversely, when a service provider is no longer available, the associated agent is stopped. As a matter of fact, autonomous agents seem to be suitable for complex functions or services that they perform directly by acting on the required resources and tools that, in turn, are not managed by any agent. Autonomous agents thus possess a complex behavior and communicate with each other in order to reach their objective. Differently from federation based architectures, communication here is not diverted through gateway agents; rather, agents can autonomously use directory resources, like yellow pages, to locate the appropriate interlocutor. Because, in principle, communication can take place among any subset of agents (differently from hierarchical- and federation-based architectures), in this case a reduced number of autonomous agents seems more fitting. Thus, the agents could represent the main actors performing quite complex tasks in the workflows that make up the business processes of an MS.

In addition, a mix of the preceding kinds of architectures (often referred to as a hybrid architecture [43]) can be followed in an MAS design approach.

The second important aspect characterizing an MAS pertains to the role assigned to the agents or, conversely, to the MS elements with which the agents are associated. Basically, the agents can be introduced to provide some service or to perform some action, or to monitor and manage the life cycle of an entity. Agents may be assigned to elements whose existence is bounded to an information world, such as procedures, functions, algorithms, data collectors, etc.; alternatively, they may be introduced to provide decision autonomy, i.e., self-consciousness, to physical elements of the MS such as products, orders, resources, tools, and so on. Shen and Norrie [42] and Shen [43] refer to this aspect as agent encapsulation, and the two alternatives mentioned previously are known respectively as *functional* and *physical decomposition*. Whenever some of the agents in an MAS are associated with functions, such as planning, and some with physical entities, such as tools, a *mixed decomposition* approach is utilized.

An agent can be generally formalized as a mapping, $\Phi\colon S \times E \times H \to A$, between a triple representing the state of the agent S, its perception of the environment, E, the history of its preceding actions, H, and the possible actions, A, that the agent can perform. The inner architecture of an agent characterizes the way in which the mapping Φ is implemented. Wooldridge [45] has classified agents into *logic-based*, *reactive*, *belief–desire–intention* (BDI), and *layered* agents according to the kind of processing they adopt to select the next action to perform. Simple agents, e.g., for monitoring physical devices, can react to a change in the environment on the basis of a procedure that simply selects the action corresponding to a specific state and input from a bidimensional look-up table. Logic-based agents, on the other hand, act as the result of a symbolic reasoning, while BDI agents must first identify the goal they seek to achieve, being in a certain state and receiving a certain input, and then use their knowledge to reach it. Layered inner architectures reflect the possibility of building hybrid types of agents whose behavior derives from the interaction among reactive- and reasoning-based modules.

Choosing the kinds of architecture, agent roles, and types to implement the most appropriate and effective MAS for P&S in manufacturing is not a simple matter. A number of valid solutions may exist, depending on the particular requirements one needs to satisfy. As a guideline for this choice, three classes of models for P&S are distinguished (even if such a classification cannot be considered clear-cut). In particular, P&S systems can be modeled by MASs as:

- An auction process among hybrid (i.e., physical and functional) encapsulated agents
- A cooperation process among agents associated with highly aggregated manufacturing facilities
- A hierarchical decomposition process among functional encapsulated agents

Planning and Scheduling MAS Approaches Based on an Auction Process

In this first model, planning corresponds to the ask-for-bid and subsequent selection of proposal phases of the auction process. In general, planning is so strictly integrated to scheduling in an MAS that a reader familiar with classic planning procedures might have the impression that planning is merely a preliminary step for scheduling. Actually, this is exactly the case for this first class of models. The auction process class of models is likely the one with the larger number of applications because it represents a "natural" way of distributing the decision capability by means of a society generally made up of synthetic social agents.

The main concepts characterizing the auction-based models can be summarized in two outstanding approaches: the Metamorph II [46] and the AARIA [47] projects. Although they propose different solutions, each provides important insights into the main characteristics presented in general by MAS approaches to P&S in manufacturing. Both models are supply chain oriented because they seek to integrate the P&S of an MS with that of the other entities involved in the supply chain, particularly the supplier or partner enterprises. They are representative of the two main alternative architectures usually adopted for MASs: federation and autonomous agents. Finally, they share some common aspects that can stimulate discussion about the appropriateness of an MAS approach to P&S in MS.

The agent modeling approaches described in these two projects exploit two of the main features of an MAS applied to P&S problems:

- Decomposition, i.e., the distribution of decision responsibility to local decisional entities
- Negotiation, which makes decisions emerge from an auction process

In other words, decomposition subdivides each single MS into a number of entities of different classes, possibly at different hierarchical levels, and then makes such entities interoperate in a sort of structured market. Decomposition is usually hybrid because the execution of logical functions and the management of physical components of the MS are assigned to agents. Separate MSs are thus integrated, making their own agents communicate at some defined homogeneous level (for example, consider the enterprise mediators in the Metamorph II system described in Chapter 6).

The contract net protocol (CNP) [48] and its modified versions are the most common negotiation protocols. A brief recent summary of its state of the art can be found in Shen [43]. In CNP, each manager agent with tasks to be assigned decides to subcontract them, broadcasting an offer and waiting for other contractor agents to reply with their own bids. The manager retains the best offers arriving within a certain elapse of time and allocates its contracts to one or more contractors. A manager agent is responsible for coordinating task allocation, providing dynamic allocation and natural load balancing. Although the basic CNP is quite simple and can be efficient, when the number of agents is large, the throughput of messages on the network may increase and produce undesirable effects, namely, network congestion and a waste of CPU time to perform bidding instead of using it to accomplish the required tasks. Improvements to the basic CNP have been proposed to overcome these problems; Shen [43] recently summarized them as:

- Multicasting, rather than broadcasting offers to a limited number of agents
- Anticipating offers, i.e., when contractors send bids in advance
- Varying the time when commitment is decided
- Allowing the breach of commitments
- Allowing coalition, i.e., when several agents can answer as a group
- Introducing priorities for solving tasks

Planning and Scheduling MAS Approaches Based on a Cooperation Process

An alternative use of MASs for P&S entails the distribution of a set of high-level functions to agents, thus making the system's behavior emerge from the cooperation among agents. This modeling alternative can be adopted when the complexity of a P&S can be distributed to several functional agents that handle the resolution of some subproblems and that should cooperate to devise a globally acceptable solution. In addition, the decomposition and cooperation approach is oriented, for example, to deal with complex MSs made up of a number of branch facilities or with clusters of enterprises, each with a limited or geographically localized production capacity, that respond as a coalition to a large and widespread flow of orders. Such systems, in fact, seek to improve overall performance by establishing a flexible level of cooperation. The accent here is not on tackling the P&S decisions at a single manufacturing facility or enterprise level, but on the way the P&S decisions can be coordinated in order to achieve a performance for the coalition that is better than the one reached by separate P&S activities.

One approach to achieve this objective is to impose some centralized structure that connects, organizes, and is in charge of decision making for all the coalition members. In truth, this may not be always acceptable, especially in the case of a set of peer enterprises, or it may not be economic and efficient, due to the difficulty of imposing a common communication standard among pre-existent entities and to the complexity of facing a global P&S problem. Thus, MASs are viewed as a flexible and effective alternative to such centralized approaches. In these settings, agents are associated with highly aggregated entities corresponding to the different functions, branch facilities, or enterprises. Thus, each single agent is responsible for the P&S of a single facility; that is, it can use any procedure or method (not necessarily agent based) to obtain a plan or simply to interface with the planning system already operating in the facility. The agents cooperate in order to improve their single plan and, consequently, the global system plan. Examples of these approaches will be reviewed again in Chapter 6.

Planning and Scheduling MAS Approaches Based on Hierarchical Decomposition Process

The third class of MAS models that can be used to cope with the P&S problems in an MS corresponds to a different way of encapsulating agents and of distributing decision responsibility; in essence, this approach may be considered a somewhat conservative manner of applying agents. The first class of models introduces agents with the purpose of going beyond the classic organization structure of MS, whereas the second class, based on a high level cooperation approach, endeavors to broaden the limited scope in which classic P&S decisions usually emerge. Both models seemingly represent a novel perspective in managing P&S activities, and they could meet with resistance in companies with consolidated and highly structured organization and practices. Therefore, the introduction of an MAS for P&S, instead of redesigning the way a company performs its business processes, must somehow adapt to the pre-existing organization.

The advantage of adopting an MAS model does not stem from the integration in space and time of all the entities involved in the manufacturing process, but on the further horizontal decomposition of the problems at the three hierarchical levels relevant to planning, scheduling, and control. In particular, business component agents are used generally here and may be responsible for planning, scheduling, and controlling the production of customer orders using even conventional procedures and methods. Nevertheless, they each take charge of the decisions about a single order or a subset of orders. Thus, several business component agents are present and characteristic of each decision level, reducing in this way the complexity through a distribution process: basically the scheduling agents receive an activity to perform from a planning agent at the upper level and decompose the activity into a set of detailed subactivities that, in turn, are assigned to lower level controlling agents. Communication among agents at the same level is possible whenever agents are allowed to subcontract part of their activities to other agents in order to meet the relevant requirements; such a possibility can be considered a local optimization policy similar to the one followed in the coordination-based model described previously.

It may be observed that the main drawback of such a fully functional decomposition is the risk of embedding the decision-making logic too heavily in the agents, instead of letting it emerge from the global behavior of the MAS. For example, the way in which the production scheduling agents operate, even if on a reduced subset of the whole problem, may be tantamount to a scheduling procedure in a classic ERP system, thus limiting the advantages of adopting an actual agent-based scheduling system.

Remarks on the MAS Approaches to Planning and Scheduling

First, consider the main differences between classic approaches to P&S and the agent-based ones thus far illustrated. Classic approaches face complexity by subdividing planning and scheduling into two horizontal hierarchical levels, which differ for the time horizon and for the aggregation considered and are longer and coarser for planning and shorter and more detailed for scheduling. Apart from on-line scheduling, P&S decisions made by classic approaches are often modified to adapt them to changes occurring outside (e.g., in customers or suppliers) and inside (e.g., failures, variations of resource availability) the MS. However, classic P&S approaches strive to find solutions by taking into account, as simultaneously as possible, all the alternative and conflicting aspects of the problems (for example, an optimal scheduling approach based on an MIP formulation explicitly includes the relevant aspects in the model constraints).

An agent-based P&S approach, by contrast, is intrinsically heuristic and separates the difficulty in a vertical or in an even further horizontal fashion; in other words, it drastically reduces the alternatives and conflicts to be evaluated and forces solutions to grow out of the composition of a sequence of distributed decisions. The overall performance of a solution implicitly derives from the performance of the single distributed decisions. Decomposition among agents calls for cooperation; conflicts are locally solved by bidding and negotiation mechanisms.

P&S activities with conventional approaches are separate activities, whereas in MASs they are strictly interwoven, as was observed in the main auction-based approaches. The first flow of information among the agents (in a federation and in an autonomous agent architecture) basically questions the production facilities about a new order, thereby defining a possible plan whose cost is evaluated; if this proposal is accepted, the resulting commitment fixes the plan into a schedule that, more precisely, corresponds to update the set of schedules of the single resources specifically involved. Such a schedule is therefore neither off-line nor defined *a priori*.

As a matter of fact, P&S by MASs tends to be intrinsically on-line because the activities to be planned and scheduled, e.g., production orders, are considered one at a time as their corresponding requests arrive, or anyway following some priority. In addition, rescheduling, i.e., the modification of planned and scheduled activity due to the occurrence of new events, is usually included in P&S MASs. Classic P&S approaches depend on the availability of reliable and updated information about the demand, lead times, and state of the MS. P&S MASs work with continuously updated information because they are strongly linked with the various sources of these data present inside the MS and in the supply chain.

In the first class of approaches, the accent is placed on negotiation; in these cases, a society of agents closely related to physical and logical elements of an MS is defined and the global system behavior emerges from the social interactions among agents in a sort of resource marketplace. The second class progressively abandons such an "anthropomorphic" point of view, relying on cooperation among higher level functional entities or on the further decomposition of the classic organization layers to provide flexibility (e.g., the ability of replanning in case of unexpected problems). Which of the two is the better approach to P&S is a recurrent question that still lacks a definitive answer. The ultimate choice clearly depends on the situations at hand, on the pre-existing infrastructure of the company's information system, and on the extent to which the company's executives and managers would seek to endow a P&S system with the autonomy to make decisions and to adapt flexibly to changing situations.

AGENT-BASED APPLICATIONS FOR SCHEDULING AND CONTROL IN MANUFACTURING

In this section, the main characteristics of MAS application to S&C in manufacturing are reviewed. As discussed in the section dealing with planning, scheduling, and control in manufacturing from an integrated perspective, it seems more appropriate to consider S&C as an integrated phase instead of separating these activities as in conventional approaches to MS management. In particular, S&C activities are strictly related whenever scheduling is assumed as an on-line activity that is not naturally or reasonably separate from control. On the other hand, a control activity without an input schedule to be followed — one that often must be defined or revised on-line — would be worthless.

The function of the shop floor control system (SFCS) is to execute the planned schedule while monitoring and controlling the production processes. In fact, a schedule determines the nominal assignment, sequencing, and timetabling of the production activities on the available resources, disregarding a number of details that, on the other hand, are necessary to actually carry out the plan. Thus, the function of SFCS is to take into account production details, from the management of movements of pallets or lots in the shop floor to the management of the tools in the manufacturing cells and the download of the correct program part to perform the operations on the numerically controlled machines.

Recent trends in manufacturing highlight the increasing importance of four main requirements for SFCS to provide the system with responsiveness and agility, namely, reconfigurability, efficiency, reactivity, and robustness. This in turn has underlined the utter inadequacy of centralized control architectures and the poor flexibility of strictly hierarchical ones, shifting the focus toward the advantages of strongly decentralizing the control capabilities. The answer to the preceding requirements is represented by heterarchical architectures, i.e., control systems made up of a number of local controllers able to decide autonomously and to cooperate in order to achieve global objectives. Thus, it becomes apparent that multiagent and/or holonic architectures can play a pivotal role in this field.

Especially when focusing on S&C, the similarity between MASs and holonic manufacturing systems is very close. In many cases, agents can be viewed as "practically" equivalent to holons from an application standpoint. Specifically, bottom level agents that directly control a physical device are often the software components of the holons introduced to decentralize the control program at the lowest manufacturing entity level (i.e., work centers, cells, robots, transport devices).

In other cases, agents are regarded as software components working in soft real time, while holons are hardware/software components working in hard real time. In all cases, agents and holons provide each entity with the autonomy of management — for example, the possibility to decide the activation of set-ups or maintenance actions on the basis of information about the state of the manufacturing activities, production requirements communicated with the schedule, and state of the device that the agent governs. Thus, no detailed commands are needed from a central or higher level controller, but only the aggregate information defining the production plan. In this decentralized framework, however, information must be exchanged among the autonomous entities associated with the resources involved (such as cells, AGVs, tools, and so on) in order to coordinate their actions and thus to complete the production of lots. This need leads to the definition of MAS/holonic systems in which the distributed autonomous decision entities are organized in some architecture, share a communication framework, and can use specifically defined protocols to interact.

As argued by many researchers in the field and underlined by McFarlane and Bussmann in a recent survey [49], distributed systems are the appropriate answer to satisfy today's requirements for production planning and control. In their analysis, the authors focus on holonic manufacturing systems and review the key features characterizing many applications. In particular, they recall that most introduce a novel distributed approach for S&C, but still within the conventional hierarchical framework for planning and control defined by the MESA-11 architecture [2]. Moreover, they admit that a genuine application of the holonic paradigm is still not yet widely adopted.

From the analysis of these authors [49], another point emerges that warrants some discussion: some work still must be done on the definition of reliable coordination protocols before a complete and effective decentralized approach based on an autonomous agent architecture covering the layers from planning to control can be considered. MASs made up of autonomous agents do not seem to be mature enough to guarantee a stable and coherent system behavior that can be forecasted as well. Thus, a hierarchical organization of autonomous entities can expedite the cooperation, stability, and robustness of the system, as well as reduce the communication load. On the other hand, the higher the level of decentralization is, the greater the capability is of the system to adapt to changes — that is, to be responsive and reconfigurable.

The advantage gained with MASs is clearly high modularity and reconfigurability, because shop floor resources can be added or updated simply by inserting new relevant agents or by updating the agents' behavior; in this way flexibility can be guaranteed, as can self-reconfigurability capacity that allows the agents to overcome a certain range of fault conditions. The pitfall here is that the high degree of autonomy granted to the single manufacturing entities through their "agentized" brain or by embedding them in holons may reveal global system behaviors that are not predefined and that may prove difficult to forecast. Entities may be in conflict and, without predefined social rules, this can lead to deterioration of system performance, as in the case of deadlocks. MASs thus require coordination, i.e., agents must communicate and interact to cooperate and resolve possible conflicts. Regarding this aspect, McFarlane and Bussmann [49] noted that most of the applications they reviewed use contract net-based interaction protocols, i.e., protocols that are oriented to negotiation rather than cooperation.

The communication capability requirement for agent coordination, especially in the shop floor control setting, calls for a high-performance communication system, that is, a local network with real-time responsiveness and an adequate band capability. In this connection, one of the possible difficulties in introducing multiagent-based SFCSs is the need to integrate them with the different and heterogeneous physical devices making up a typical manufacturing shop floor. Although efforts have been made toward standardization of communication protocols, the interaction with different manufacturing resources, from robots and numerically controlled machines to AGVs, often requires an interface among different legacy controllers, each operating with a different language that may centralize the governing of multiple devices.

The distributed holonic model may thus represent an alternative to the traditional centralization of functions in PLCs. However, theoretic models are usually difficult to implement effectively in the typical shop floor, which is oriented toward practical applications that are strongly constrained by standards and where PLCs dominate in a traditional well-assessed architecture. The IEC 61499 standard [50, 51] appears to address this problem with a bottom-up approach starting from the shop floor, following the rules of control systems, and preparing an effective "landing" of holons and agents at the shop floor level.

This standard deals with the application of function blocks (FBs) in distributed industrial-process measurement and control systems (IPMCSs) in the so-called low-level control (LLC) domain of a holonic system. LLC addresses traditional control/automation functions by means of FBs operating in hard real time (response times: 10 μs to 100 ms) and communications in short messages (lengths: 1 bit to 100 bytes) [50]. LLC is integrated with the high-level control (HLC) domain that addresses traditional scheduling operations implemented by negotiation/coordination protocols achieved through agents working in soft real time (100 ms to 10 s) and communications in longer messages (100 bytes to 4 kbytes). IEC 61499 can thus be deemed a leading standard for LLC, just as FIPA [52] is for HLC.

The major challenges in taking the application of FBs from classical PLCs to distributed IPMCSs derives from the fact that FBs are traditionally regarded as elements of programs in a centralized controller, while in IEC 61499 they are elements of distributed applications in a decentralized control system. The IEC 61499 model of a system consists of a number of devices, which can communicate with each other over communication networks; can interface with physical processes, equipment, or machines; and can serve as platforms for the execution of distributed applications [50]. The application model includes an FB network interconnected by flows of events and data over event connections and data connections, respectively. The elements of these applications are in principle distributable among multiple devices. FBs, in turn, are considered to be instances of FB types (classes), which are specified in formal declarations using the means defined in IEC 61499-1. A device may consist of multiple resources, which may share communication and process interfaces. Each resource may contain local applications, or the local parts of distributed applications, and may provide a platform for scheduling the execution of algorithms in FBs and for mapping underlying operating system services.

The IEC 61499 standard allows encapsulation and reuse of control algorithms by end users and, in a certain way, lays solid bases for manufacturing agility. Modern industrial processes require even higher degrees of physical reconfigurability and must undergo constant "metamorphic" [39] transformations in order to accommodate the frequent changes in product mix and volume described in Chapter 1. Consequently, in order to achieve peak performance at the control level, it will ultimately be desirable to have intelligently self-reconfiguring devices that can accomplish task assignments decided by negotiation with other holons. The IEC 61499 standard is not far from achieving this goal.

PLANNING, SCHEDULING, AND CONTROL IN PS-BIKES

Chapter 2 introduced the characteristics of the PS-Bikes company, an imaginary enterprise that manufactures bicycles. In that chapter, the issues pertinent to making PS-Bikes flexible in processing orders from an e-commerce site through the introduction of an MAS solution were discussed. However, some questions were purposefully left open, especially regarding the possible integration of the new northern plant with the company's older one, the introduction of an agent-based MES in the new plant, and the adoption of an agent-based/holonic S&C system in both production facilities. This section endeavors to answer these questions.

Toward the New Northern PS-Bikes Plant Integration

An MAS for workflow management that handles new orders coming from PS-Bikes' e-commerce site was defined in Chapter 2; thus, a natural starting point for the integration of the new northern plant in order to exploit its production capacity is to share the workload of the orders received from the Web site with this plant. This choice is likely to be preferred by PS-Bikes' executives, who are seeking to replace the consolidated procedures gradually, based on the conventional layered planning system.

To reach this goal, the MAS architecture proposed in the section in Chapter 2 titled "from problems to agents at PS-Bikes" is replicated in the new northern plant, introducing the three sales, production, and purchase agencies into the plant information system. Then, following the cooperation paradigm discussed in the section on agent-based applications in manufacturing planning and scheduling, coordination between the two separate sets of agencies is achieved by introducing a new MAKE-IT agent with a sales supervisor role. The supervisor must assume responsibility for the tasks of monitoring the arrival of new orders from the Web and announcing them to the sales agencies of the two plants. Behaving similarly to a shop floor mediator described in Shen and Norrie [46], the supervisor waits for the offers from the sales agencies of the two plants and then compares them in terms of delivery time and total cost, taking into account as well the utilization level of the plants. The supervisor can directly select one of the two offers if it is clearly dominant or if it needs to keep the utilization levels close to the previously fixed thresholds; it can propose both alternatives to the customer (e.g., via the e-commerce application), directly asking their preference. The resulting commitment finally inserts the order into the production plan of the selected branch.

MAS Model for Planning and Scheduling in the New PS-Bikes Plant

Now consider how to select an appropriate model to introduce an agent-based/holonic planning, scheduling, and control system in the new northern plant. The choice should stem from considerations about the nature of the production, i.e., of demand and processes, and about the company's objectives. Customer demand for PS-Bikes' products is not constant and is clearly seasonal, but in any case it has always been considered quite stable. The major "positive" disturbances come from the unforeseeable flow of very small orders from the soon-to-be launched e-commerce site. In addition, at the northern plant PS-Bikes' executives would like to experiment with new product lines, introduction of new machines, and the possibility of ultimately offering product customizations.

Generally speaking, an information system consists of two main components: (1) a passive one corresponding to the hardware and software infrastructures for data management (usually implemented by means of RDBMS or object-oriented databases) and for communication (usually a LAN with a TCP/IP-based intranet); and (2) an active one represented by the business processes, including logical functions, such as off-line planning and scheduling, and on-line operational procedures such as monitoring and control. The choice for a multiagent or holonic manufacturing system basically involves the way in which an active information system component is modeled and implemented — that is, the way of establishing to which autonomous entities, relationships, and behaviors the execution of business processes is assigned. From the discussion of the previous sections, the choice lies between two extreme alternatives: a flat architecture made up of autonomous peer agents, or an agent hierarchy (or oligarchy) in which subsets of agents are in charge of supervising and driving the execution of the business processes by influencing the behavior of lower level agents.

Some further remarks may now be helpful. The choice of introducing a hierarchy of agents that functionally decomposes the conventional layered ERP-MES-SFCS architecture, assigning to agents the classic planning scheduling and control functions, does not seem to present a meaningful change from a centralized system, apart from the clear benefits of decentralization at the physical control level. In fact, although the decision about autonomous or hierarchical agent architectures for the execution level is still open to debate, the presence of autonomous agents (holons) at the control level through the introduction of a supervisor agent to deal only with situations requiring a wider degree of coordination has been recognized as a necessity [53]. Thus, taking into account the considerations reported in the section on agent-based applications for scheduling and control in manufacturing, and observing that the high-priority objectives for PS-Bikes are (1) to satisfy as much as possible the commitment to customers in terms of delivery times and (2) to provide the market with the perception of company reliability, a hybrid, or semihierarchical, solution seems preferable.

Such a decision provides only a general orientation because many degrees of freedom still remain to be analyzed. The MAS architecture can be a federation, in which several mediator agents at different levels are responsible for defining production plans and schedules and interact with customers on the one side and with suppliers on the other in an SCM setting. In particular, the architecture reported in Figure 3.6 can be adopted.

In this figure, rectangular boxes represent mediator agents with higher level tasks, whereas boxes with rounded corners stand for agencies, i.e., sets of agents with a given homogeneous operational function. The MAS, equipped to interconnect the plant in the PS-Bikes' supply chain, at its outset is linked to the company's administrative department, which collects

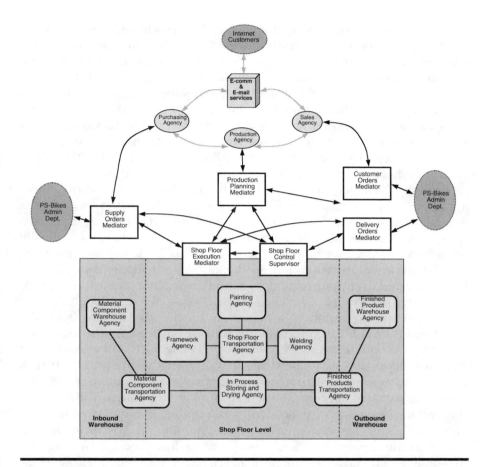

Figure 3.6   Proposed MAS Architecture for Planning, Scheduling, and Control at PS-Bikes

and forwards conventional customer orders as well as manages relations with suppliers. The agencies introduced to cope with orders from the e-commerce site, denoted with ovals in Figure 3.6, are also integrated in the architecture, making them communicate with the mediators managing the relations with customers and suppliers and with the one managing the production planning. The activity of P&S emerges from an auction-based process similar to the one described in the section on planning and scheduling MAS approaches based on an auction process: customer requests are sent to the production planning mediator (PPM), which inserts them into a tentative plan according to estimated information about the plant's future available capacity and begins a request for a bid cycle, communicating the characteristics of the requests and the relevant tentative plan to the shop floor execution mediator (SFEM). The SFEM decomposes a request in its production operation specifications, which are then delivered according to the correct sequence to the appropriate shop floor resources through their relevant agencies.

The feedback from the agencies represents the proposals (in terms of time and cost) for the execution of the operations. The SFEM selects the ones that allow satisfying the request at the lower cost, particularly with minimum deviation from the required due dates, and updates the tentative schedule. The SFEM then sends the feasible proposal to the PPM, which in turn updates the plan with the request communicated to the customer through the customer order mediator. The possible customer commitment about the proposal fixes the plan and the schedule for the order that is communicated to the shop floor resources involved through the shop floor control supervisor (SFCS), a mediator agent in charge of actually driving the on-line execution of schedule. The SFCS continuously monitors the execution of operation and, in the case of variations from the plan or unexpected events, first attempts a local rescheduling or asks the SFEM to redefine (part of) the schedule on-line. Note that even though this has not been explicitly reported in Figure 3.6 for the sake of simplicity, the SFEM and the SFCS can communicate with all of the shop floor agencies.

Now concentrate on the way in which a schedule can be defined for PS-Bikes' shop floor resources. As an example, the framework area is considered. As previously outlined, the agents populating the framework agency have two main functions: contracting of the requests for new jobs (orders) to define a plan (a schedule) and on-line scheduling of the jobs according to the planned priority while controlling their execution. The agents in the framework agency perform the first function by adopting the on-line scheduling mechanism based on a negotiation protocol as described later. To provide production flexibility, the planned schedule actually identifies job priorities and desired delivery times; however, final scheduling decisions are left to the execution level so that new urgent orders and small Web orders have the chance to compete to be processed against the major flow of wholesalers' orders. The second function of the agents in the framework agency, corresponding to the S&C function, is that of on-line implementation and monitoring of the schedule's execution and possibly on-line rescheduling by starting a new negotiation cycle.

The laser cutters operating in the framework area must perform the first task of bicycle production: cutting the tubes according to the desired specifications for the frames. Jobs may correspond to production batches of different sizes, thus giving rise to generally different processing time requirements for the cutting tasks. Thus, the framework area is composed of a set of parallel machines, the laser cutters, that must process a set of jobs arriving over time and characterized by a processing requirement, a priority, a due date, and a couple of penalties for the job's early or late completion. The reason for these penalties is that of discouraging an excessively early execution of a job to force a just-in-time policy in material supply, as well as an excessively overdue execution to keep customer satisfaction high with on-time deliveries. The on-line scheduling mechanism adopted for the framework area represents an adaptation of the multiagent scheduling system proposed by Gozzi et al. [54, 55] and is based on the following five types of agent:

- One job-agent generator agent (JGA)
- One machine-agent generator agent (MGA)
- One contract coordinator agent (CCA)
- A set of job agents, including a job agent, JA, for each job
- A set of machine-agents, including a machine-agent, MA, for each machine

Three types of agents, namely, the two generator agents JGA and MGA, and the coordinator, CCA, are permanent and always active. In order to define a plan and schedule on the PS-Bikes shop floor, the generator agents and the coordinator agent communicate with the SFEM, respectively receiving from it the announcement of jobs to execute and the possible activations of new machines and informing it about the committed scheduling decisions. When a schedule is subsequently executed, the shop floor agents communicate with the SFCS, which oversees the monitoring of the actual manufacturing operations.

The jobs are released over time by the SFEM according to the defined plan and on the basis of the actual on-line arrivals (urgent or Internet orders); they are then delivered to the various shop floor areas in order to define a schedule for them. In particular, every time a job enters the framework area, a new JA is created by the JGA; a generic JA, j, knows the standard required processing time, p_j; the due date, d_j; and the weights for the early completion, we_j, and tardy completion, wt_j, with respect to the due date of the associated job. In addition, the JA has a budget of virtual money, assigned by the JGA on the basis of the job's priority, to purchase the service from the MAs.

On the other hand, an MA knows its actual speed (the standard speed is equal to 1); the list of assigned jobs; and a list describing its available time slots. The objective of any JA is to obtain a service for the job j from an MA so that the job is completed as close as possible to its due date, thereby avoiding penalties for early or tardy completion. Any MA tries to maximize the number of accomplished jobs and the amount of virtual money received for the service. The schedule thus emerges from the

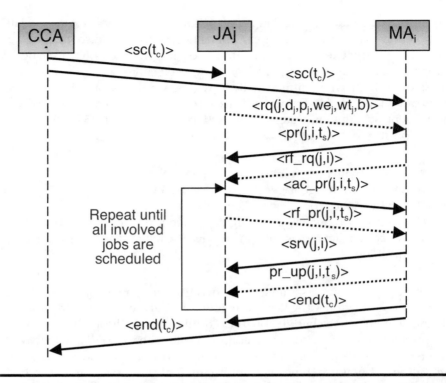

Figure 3.7   The Negotiation Protocol Among the Agents in the Framework Agency

negotiation among JAs on one side and MAs on the other, a cycle that follows a contract net-like schema as shown in Figure 3.7.

The negotiation cycles are driven by the CCA, which sends a start cycle signal (`<sc(tc)>`) to the JAs and MAs informing them of the current time, t_c. The JAs decide, depending on t_c and their due dates, whether to send a service request message to the MAs or to wait for a later cycle. Such a decision is made by each JA j on the basis of a probability of requesting service at time t_c, $p(t_c, d_j)$, which is empirically assigned according a request probability selection function (RPSF), as depicted in Figure 3.8.

The RPSF is a function of the float f of the job j, i.e., of the time lapse from the current time to the job's ideal starting time (so that job j is completed at its due date d_j), and depends on several parameters representing time thresholds (T_1, T_2, T_3) and probability levels (P_1, P_2). Different choices for the parameters allow representing different possible inclinations of the JAs to request service early or to wait and must be tuned to the characteristics of the jobs a company would schedule. In fact, having selected a probability value, the decision whether to request a service during the current negotiation cycle is randomly determined, thus giving

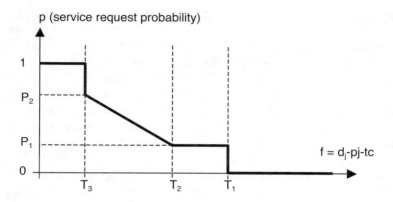

Figure 3.8 Request Probability Selection Function of a JA

low-priority orders additional chances of negotiation. Whenever a JA sends a service request message (`<rq(j,dj,pj,wej,wtj,b)>`) to the MAs, it specifies its relevant data (due date, d_j, processing time, p_j, and weights for the earliness and tardiness, we_j, wt_j) and a bid (b) that is determined with a bid selection function (BSF) analogous in shape and parameters to the RPSF.

When a service request is received, an MA first determines whether to accept or to refuse it with a random choice based on a probability value obtained from an acceptance probability selection function (APSF); if it accepts a request, it prepares and sends a proposal to the JA. Even in this case, the APSF empirically links the acceptance probability to the job's float and bid and expresses the propensity of an MA to accept the requests (e.g., early requests with large float can be discouraged, and larger bids can be favored).

Even the APSF is characterized by a number of parameters that must be tuned in order to tailor the scheduling heuristic to the PS-Bikes' context. If MA *i* rejects a request from JA *j*, it then sends a refusal message (`rf_rq(j,i)`) to the JA. Otherwise, if the request is accepted, the MA searches its list of available time slots for an interval in which it can accomplish the job incurring the lowest (weighted) penalty and then replies with a proposal message (`pr(j,i,t_s)`, where t_s is the proposed starting time). The JA can in turn evaluate the proposals from the MAs, accepting the best one (answering an `ac_pr(j,i,t_s)` message) and refusing the others (with an `rf_pr(j,i,t_s)` message). After having received one or more acceptance messages from the JAs, the MA sorts them according to associated bids and serves the JA with the greatest bid as agreed, updating the time availability list and sending a commitment message (`srv(j,i)`).

The MAs then consider the additional requests in the bid order and commit the proposed interval to them whenever possible (i.e., if the intervals are still available after the previous commitment), or reply to JAs with a proposal with an updated interval (pr_up((j,i,t'$_s$))), thus starting a new iteration of negotiations. Whenever the MAs update a proposal, the JAs again evaluate them and select the new best candidate until a commitment with an MA is reached. This negotiation process terminates after a finite number of steps because, at each iteration, at least one JA certainly will obtain the service and will leave the competition.

The schedule is watched step by step by the CCA; in fact, the CCA records all the successful contracts and updates the global partial schedule. Its main role, however, is to act as time manager, communicating the current time to the agents and interrupting excessively long iterations according to a prefixed maximum duration that forces the negotiation cycle to evolve expediently in order to satisfy the possible real time requirements.

Summing up, the agent-based scheduling mechanism described is essentially a negotiation similar to contract net, with a randomization of the JA and MA decisions. The relevant selection functions do not deterministically force the choice of agents, but only point out a tendency. What is the underlying rationale? From an on-line scheduling algorithm standpoint, stochastic compared to deterministic choices of agents allow for some unforeseeable events (for example, selecting the request for service with the highest float or with the lowest bid), thus preventing the systematic assignment of resources to a homogeneous kind of lot, and maintaining a balance in the average performance of lots. In the case of PS-Bikes, this means handling the new heterogeneous flow of orders without fixing *a priori* a production mix that, given the uncertainty characterizing Web-based orders, can lead to poor performances. As already observed, the selection functions parameters should be tuned to adapt the scheduling mechanism to the specificity of PS-Bikes' production. Actually, the best solution now on the horizon is that in which the agent system is given a learning capability that enables the (possible dynamic) tuning of parameters.

The performances of the on-line agent-based scheduler with respect to a set of randomly generated test problems have been evaluated in Gozzi et al. [55]. In particular, the average competitiveness ratio (ACR), i.e.,

$$E\left[\frac{C(S, I)}{C(S_{opt}, I)}\right]$$

has been experimentally estimated, where $C(S,I)$ is the objective function for a solution S provided by the agent-based scheduler for a problem instance I, with S_{opt} representing the optimal solution for the same instance.

The problem instances with a number of jobs, $N \in \{5, 10, 15, 20, 25, 30, 40, 50\}$, and a number of machines, $M \in \{1, 2, 3, 4, 5\}$, have been used for this test. Numbers are intentionally kept small in order to obtain the optimal solution of the problems, which are needed to estimate the ACR by means of an MIP formulation (derived in a straightforward fashion from the method in Balakrishnan et al. [56]) within an acceptable computation time. In fact, the optimal solution was successfully computed only for a subset of randomly generated problems (about 62% out of 217 problems). Nevertheless, the value obtained for the ACR, equal to 1.52, seems to be telling of the promise of the agent-based approach.

Scheduling and Control in the New PS-Bikes Plant

Consider again the framework area of the PS-Bikes' shop floor. The monitoring and controlling of the production is performed by JAs, MAs, and SFCS. The JAs live until the job is not executed; every variation in the future schedule of the job activates the agent that, if prompted, can start again a negotiation cycle to obtain a new feasible service slot or to improve the job performance. The MAs manage the machines' local schedules (and the machines' agendas), updating them in the case of deviations of actual machining times from the computed processing times. In such cases, e.g., an unrecoverable machine breakdown, a type of message different from the ones used in the scheduling negotiation is sent to JAs, the CCA, and the SFCS in order to let those agents reconsider a new machine in order to be served. Finally, the SFCS follows the whole schedule's execution continuously by means of the different agencies in the PS-Bikes shop floor, controlling the correct advancement of the production by announcing a job to an agency only when the preceding operations on the same job by the previous agency have been completed.

All the previous processes work in soft real time and can be considered HLC (see the section on agent-based applications for scheduling and control in manufacturing). To exploit its function, HLC must be integrated with LLC; this can be implemented in a conventional, PLC-based, hierarchical/centralized way or according to a holonic manufacturing system design. Specifically, in the new plant, a holonic manufacturing system is evaluated for implementation, and the recommendation is to follow the holonic system specifications of IEC 61499 [50, 51]. The use of IEC 61499 in the design of a distributed control of a new LLC for the new PS-Bikes plant will proceed according to two phases [51]. First, a *functional design phase* is performed in which process engineers analyze the physical plant design, for example, using ANSI-ISA 5.1 [56] diagrams representing a series of blocks outlining the main software components and their primary interconnections. A *functional distribution phase* then defines the distribution of control functionality on to processing resources. The IEC 61499 standard provides models and concepts to define the distribution of functionality in interconnected IEC 61499 function blocks [51].

During the functional distribution phase, system engineers should complete the detailed design by mapping the software requirements on each block. Some blocks will need to be designed "ad hoc" for that system application; in other cases, blocks can be reused with standard instrumentation and controllers. Moreover, during this phase, other blocks will be defined [51], for example, communication blocks, such as server and client blocks, which can be used to formalize the exchange of data between blocks in different physical processing resources, and interface blocks that provide interfaces with the processing resource infrastructure. Thus, it is clear that in the design of the new holonic-based PS-Bikes manufacturing plant, software design is as important as hardware design.

A practical and detailed example showing the application of the elements of IEC 61499 can be found in Christensen [50], in which an actuator moves a workpiece along a slide in the "forward" direction at a given velocity, VF, and in the "reverse" direction at a given velocity, VR. In the PS-Bikes design of the new plant, several of these actuators are present, for example, in the laser cutters.

CONCLUSIONS

To conclude, some remarks on MAS applications to manufacturing planning, scheduling, and control problems can be made. Throughout the chapter, two main features emerged for MAS approaches:

■ MASs represent a means to implement decentralization in MS, i.e., to decompose the decision problems relevant to manufacturing production into a set of less complex subproblems tackled by one or more decision entities.

■ MASs entail a flexibility that provides the preceding decision entities with autonomy and proactiveness, if needed. Regardless of the design choices adopted for an MAS to adapt it to a specific MS application, the MAS approach seems the best current technological solution to introduce agility in manufacturing.

In particular, recall the MS agility requirements, from the bottom up:

■ Agility requires a production control model, which is distributed and highly reconfigurable, in order to respond to the modern requirements of (1) decentralizing the control procedures, currently embedded in PLC, directly on manufacturing workstations; and (2i) introducing the current facilities of high-level software (such as the reusability in object-oriented programming) into control software components. Agent or holonic systems satisfy both requirements because they clearly enable the decentralization of the control of single shop floor devices through the assigning of each to a single agent or holon that, in turn, is implemented according to recent software engineering specifications and allows simple software component reconfigurability and reuse.

■ Agile scheduling demands the ability to alter decisions dynamically about production activities in order to react rapidly to changes in the supply chain; on-line scheduling and rescheduling facilities will be an increasingly compulsory component of next-generation MES. Agent-based scheduling systems are able to implement the preceding properties; in particular, complex scheduling decisions can be handled by systems made up of synthetic social agents, which again are a means to decompose the problems, whose feasible solution generally emerges from a negotiation-based protocol.

■ Off-line plans will become progressively less significant in an agile manufacturing context because they are based on coarse and often not updated information. Planning by MAS is strictly related to scheduling and, in particular, MAS architectures for P&S aim to guarantee interoperation of all the relevant entities in the supply chain; this allows a direct flow of information in the chain, making planning decisions fresher and more reliable.

In addition, an MAS can be introduced to model business processes along the supply chain: in particular, business component agents are suitable to model and implement distributed workflow management systems.

In conclusion, agent-based approaches offer promising solutions to manufacturing planning, scheduling, and control problems and provide many advantages when a dynamic, distributed, and reconfigurable architecture is needed, as in the case of agile manufacturing. It is the authors' opinion, as well as those reported in other outstanding reviews (for example, Shen [43]), that in order to attain this goal gradually, combinations of agent-based and conventional approaches may also be expedient. Again quoting Shen [43], it could be concluded that "whether implementations of MASs realize these potential advantages will depend on selecting a suitable system architecture [preferably assessed within a standard] as well as on effective mechanisms and protocols for communication, cooperation, coordination, negotiation, adaptation, ..." as well as knowledge encapsulation, reusability, and other capabilities that can be achieved by integrating the research results of several disciplines, including operations research, software engineering, game theory, etc. After the review in Chapter 6 of the most relevant real-world applications of agent-based manufacturing, these challenging research aspects will be further explored in Chapter 7 as a conclusion to this book.

REFERENCES

1. McClellan, M., *Applying Manufacturing Execution Systems*, St. Lucie Press, Boca Raton, FL, 1997.
2. MESA International, Controls definition and MES to controls data flow possibilities, White Paper n.3, 1995.
3. ANSI/ISA-95.00.01-2000, Enterprise-control system integration part 1: models and terminology, ANSI/ISA-95.00.02-2001, Enterprise-control system integration part 2: object model attributes, available at http://www.isa.org/.
4. Thomas, L.J. and McClain, J.O., Overview of production planning, in *Logistics of Production and Inventory, Handbooks in Operations Research and Management Science*, Graves, S., Rinnooy Kan, A.H.G., and Zipkin, P.H., Eds,, North–Holland, Amsterdam, 4, 1993.
5. Nahmias, S., *Production and Operation Analysis*, 3rd ed., Irwin, Burr Ridge, IL, 1997.
6. Hadley, G. and Whitin, T.M., *Analysis of Inventory Systems*. Prentice Hall, Englewood Cliffs, NJ, 1963.
7. Erlenkotter, D., Ford Whitman Harris and the economic order quantity model, *Operations Res.*, 38, 937, 1990.
8. Wagner, H.M. and Whitin, T.M., Dynamic version of the economic lot size model, *Manage. Sci.*, 5, 89, 1958.
9. Silver, E.A. and Meal, H.C., A heuristic for selecting lot size quantities for the case of a deterministic time-varying demand rate and discrete opportunities for replenishment, *Prod. Invent. Manage.*, 14, 64, 1973.
10. Shapiro, J.F., Mathematical programming models and methods for production planning and scheduling, in *Logistics of Production and Inventory, Handbooks in Operations Research and Management Science*, Graves, S., Rinnooy Kan, A.H.G., and Zipkin, P.H., Eds, North–Holland, Amsterdam, 4, 1993.
11. Wolsey, L.A., MIP modelling of changeovers in production planning and scheduling problems, *Eur. J. Operational Res.*, 99, 154, 1997.
12. Drexl, A. and Kimms, A., Lot sizing and scheduling — survey and extensions, *Eur. J. Operational Res.*, 99, 221, 1997.
13. Clark, A.R., Approximate combinatorial optimization models for large-scale production lot sizing and scheduling with sequence-dependent setup times, in *Proc. IV ALIO/EURO Workshop Appl. Combinatorial Optimization*, Pucón, Chile, November 2002.
14. Khouja, M., Michalewicz, Z., and Wilmot, M., The use of genetic algorithms to solve the economic lot size scheduling problem, *Eur. J. Operational Res.*, 110, 509, 1998.
15. Kimms, A., A genetic algorithm for multilevel, multimachine lot sizing and scheduling, *Computers Operations Res.*, 26, 829, 1999.
16. Disney, S.M., Naim, M.N., and Towill, D.R., Genetic algorithm optimisation of a class of inventory control systems, *Int. J. Prod. Econ.*, 68, 259, 2000.

17. Crauwels, H.A.J., Potts, C.N., and Van Wassenhove, L.N., Local search heuristics for single-machine scheduling with batching to minimize the number of late jobs, *Eur. J. Operational Res.*, 90, 200, 1996.

18. Meyr, H., Simultaneous lotsizing and scheduling by combining local search with dual reoptimization, *Eur. J. Operational Res*, 120, 311, 2000.

19. Kim, J. and Kim, Y., Simulated annealing and genetic algorithms for scheduling products with multilevel product structure, *Computers Operations Res.*, 23, 857, 1996.

20. Barbarosoglu, G. and Özdamar, L., Analysis of solution space-dependent performance of simulated annealing: the case of the multilevel capacitated lot sizing problem, *Computers Operations Res.*, 27, 895, 2000.

21. Maes, J. and Wassenhove, L.V., Multi-item single-level capacited dynamic lotsizing heuristics: a general review, *J. Operational Res. Soc.*, 39, 991, 1988.

22. Clark, A.R. and Armentano, V.A., A heuristic for a resource-capacitated multistage lot-sizing problem with lead times, *J. Operational Res. Soc.*, 46, 1208, 1995.

23. Lawler, E.L., Lenstra, J.K, Rinnooy Kan, A.H.G., and Shmoys, D.B., Sequencing and scheduling: algorithms and complexity, in *Logistics of Production and Inventory, Handbooks in Operations Research and Management Science*, Graves, S., Rinnooy Kan, A.H.G., and Zipkin, P.H., Eds., 4, North–Holland, 445, 1993.

24. French, S., *Sequencing and Scheduling: An Introduction to the Mathematics of the Job-Shop*, Ellis Horwood Ltd., Chichester, England, 1982.

25. Baker, K.R., *Introduction to Sequencing and Scheduling*, John Wiley & Sons, New York, 1974.

26. Blazewicz, J., Ecker, K.H., Pesch, E., Schmidt, G., and Weglarz, J., *Scheduling in Computer and Manufacturing Systems*, Springer Verlag, Berlin, 1994.

27. Garey, M.R. and Johnson, D.S., *Computers and Intractability: A Guide to the Theory of NP-Completeness*, W.H. Freeman and Co., New York, 1979.

28. Pinedo, M., *Scheduling: Theory, Algorithms & Systems*, Prentice Hall, Upper Saddle, NJ, 2002.

29. Brucker, P., *Scheduling Algorithms*, Springer Verlag, Berlin, 2001.

30. Sgall, J., On-line scheduling, in *On-line Algorithm: the State of the Art, Lecture Notes in Computer Science*, Fiat, A. and Woeginger, G.J., Eds., Springer, Berlin, 1442, 196, 1998.

31. Baker, K.H. and Scudder, G.D., Sequencing with earliness and tardiness penalties: a review, *Operations Res.*, 30, 22, 1990.

32. Kanet, J.J. and Sridharan, V., Scheduling with inserted idle time: problem taxonomy and literature review, *Operations Res.*, 48, 99, 2000.

33. Panwalkar, S.S. and Iskander, W., A survey of scheduling rules, *Operation Res.*, 25, 45, 1977.

34. Ferrel, W., Jr., Sale, J., Sams, J., and Yellamraju, M., Evaluating simple scheduling rules in a mixed shop environment, *Computers Ind. Eng.*, 38, 39, 2000.

35. Graham, R.L., Bounds for certain multiprocessor anomalies, *Bell Syst. Tech. J.*, 45 1563, 1966.

36. Borodin, A. and El-Yaniv, R., *Online Computation and Competitive Analysis*, Cambridge University Press, New York, 1998.

37. Irani, S. and Karlin, A.R., Online computation, in *Approximation Algorithms for NP-Hard Problems*, Hochbaum, D.S., Ed., PWS Publishing Company, Boston, 1996.

38. Liu, X. and Zhang, W.J., Issues on the architecture of an integrated general-purpose shop floor control software system, *J. Mater. Process. Technol.*, 76, 261, 1998.

39. Balasubramanian, S., Brennan, R.W., and Norrie, D.H., An architecture for metamorphic control of holonic manufacturing systems, *Computers Ind.*, 46, 13, 2001.

40. John, K.-H. and Tiegelkamp, M., *IEC 61131-3: Programming Industrial Automation Systems. Concepts and Programming Languages, Requirements for Programming Systems, Aids to Decision-Making Tools*, Springer, New York, 2001.

41. Christensen, J.H., Function block-based, holonic systems technology, available at www.holobloc.com.

42. Shen, W. and Norrie, D.H., Agent-based systems for intelligent manufacturing: a state-of-the-art survey, *Knowledge Inf. Syst. Int. J.*, 1, 129, 1999.

43. Shen, W., Distributed manufacturing scheduling using intelligent agents, *Intelligent Syst.*, 17, 88, 2002.

44. Baker, A.D., A survey of factory control algorithms which can be implemented in a multiagent heterarchy: dispatching, scheduling and pull, *J. Manuf. Syst.*, 37, 297, 1998.

45. Wooldridge, M., Intelligent agents, in *Multiagent Systems: A Modern Approach to Distributed Artificial Intelligence*, Weiss, G., Ed., MIT Press, Cambridge, 1999.

46. Shen, W. and Norrie, H.N., An agent-based approach for dynamic manufacturing scheduling, in *Proc. Autonomous Agents'98 Workshop Agent-Based Manuf.*, Minneapolis/St.Paul, MN, 117, 1998.

47. Parunak, H.V.D., Baker, A.D., and Clark, S.J., The AARIA agent architecture: from manufacturing requirements to agent-based system design, *Integrated Computer-Aided Eng.*, 8, 45, 2001.

48. Smith, R.G., The contract net protocol: high-level communication and control in a distributed problem solver, *IEEE Trans. Computers*, C-29, 1104, 1980.

49. McFarlane, C.D. and Bussmann, S., Development in holonic production planning and control, *Int. J. Prod. Control*, 11, 522, 2000.

50. Christensen, J.H., HMS/FB architecture and its implementation, in *Agent-Based Manufacturing. Advances in the Holonic Approach*, Deen, S.M., Ed., Springer, 53, 2003.

51. Lewis R., Modelling distributed control systems using IEC 61499. Applying function blocks to distributed systems, IEEE 2001.

52. The Foundation for Intelligent Physical Agents, available at http://www.fipa.org/.

53. Rabelo, R.J., Camarinha–Matos, L.M., and Afsarmanesh, H., Multi-agent-based agile scheduling, *Robotics Autonomous Syst.*, 27, 15, 1999.

54. Gozzi, A., Paolucci, M., and Boccalatte, A., Autonomous agents applied to manufacturing scheduling problems: a negotiation-based heuristic approach, in *Multi-Agent Systems and Application II, Selected Revised Papers: 9th ECCAI-ACAI/EASSS 2001, AEMAS 2001, HoloMAS 2001, LNAI 2322*, Marik, V., Stepankova, O., Krautwurmova, H., and Luck, M., Eds., Springer Verlag, 194, 2002.

55. Gozzi, A., Paolucci, M., and Boccalatte, A., A Multi-agent approach to support dynamic scheduling decisions, in *Proc. 7th Int. Symp. Computers Commun. — ISCC2002*, Taormina, (I), 983, 2002.

56. Balakrishnan, N., Kanet, J.J., and Sridharan, V., Early/tardy scheduling with sequence dependent setups on uniform parallel machines, *Computers Operations Res.*, 26, 127, 1999.
57. ISA-5.1-1984 - (R1992): Instrumentation symbols and identification available at www.isa.org.

4

AGENT-BASED SIMULATION

Manuel Gentile, Massimo Paolucci, and Roberto Sacile

Simulation is commonly used in agent research as a way to validate the design and to analyze the influence of different design alternatives in the performance of systems. This chapter deals with agent-based simulation and, in particular, with the problem of building and simulating models of manufacturing systems (or parts of them) with a prevalent multiagent-based component. The specific features of this kind of simulation model are discussed also with reference to traditional simulation approaches adopted in manufacturing and to available agent-based simulation frameworks. The chapter concludes by proposing an application of multiagent-based simulation to tackle the problem of the configuration of the multiagent scheduling system employed in the fictitious PS-Bikes company.

INTRODUCTION

This chapter deals with agent-based simulation (ABS) in the attempt to illustrate what characterizes and what differentiates a multiagent-based simulation (MABS) model from other conventional (i.e., non agent-based) simulation models. It must be underlined from the outset that we are focusing on *computer simulation* — in particular, on simulation applications in manufacturing. As a matter of fact, simulation is a very broad field that provides useful and often essential insights into a wide range of scientific and application sectors. In the last 5 years, the interest of multiagent system (MAS) and simulation scientists in agent-based simulation has grown steadily; the first international workshop on MASs and

MABSs was promoted by the International Conference on MAS (ICMAS) in Paris in 1998, and the first conference on agent-based simulation (ABS) of the Society for Modeling and Simulation (SCS) dates to 2000. Thus, assuming that the importance of simulation in supporting operational decisions in manufacturing to improve performance is not open to debate, the point to establish is whether agent-based simulation is a new, better tool compared to traditional simulation models and techniques currently used in manufacturing.

To this end, the following section briefly summarizes the meaning of simulation, specifically focusing on the kinds of computer simulation models available to support decision-making in manufacturing. Then, the third section delves into MABS models and architectures and their specific characteristics, highlighting when the application of such models is suitable. The section on platforms to support multiagent-based simulation development goes a bit more into technical aspects, considering some of the most popular tools for developing agent-based simulators and the relevant modeling design issues. The section containing an application of an MABS in the PS-Bikes company closes the chapter by inviting the reader to consider from an ABS viewpoint the agent-based scheduling model introduced in Chapter 3 for the PS-Bikes' framework area. This section, in fact, highlights a possible use in our fictitious manufacturing company of this kind of simulation model to reach peak performance.

MODELING AND SIMULATING IN MANUFACTURING

We first must establish some general points and concepts about modeling and simulating, focusing particularly on the simulation models and techniques usually adopted in manufacturing.

Simulation is the technique of imitating the behavior of a system or process (e.g., physical, economic, mechanical, social, and so on) by means of a suitably analogous artificial system or process. Two general definitions of simulation are worthy of citation: "Simulation means driving a model of a system with suitable inputs and observing the corresponding outputs" [1]; and "A simulation is the imitation of the operation of a real-world process or system over-time" [2]. In other words, simulation can be viewed as a method to analyze the behavior of real systems and the effect of exogenous interventions on such systems that modify them and/or their inputs.

The core of this method is the simulation model; in computer simulation, the model is generally a set of variables, constants, equations, functions, and input data files used and structured in a set of software modules, the simulator program, whose execution allows registering a set of output data representing the imitation of the actual system's behavior.

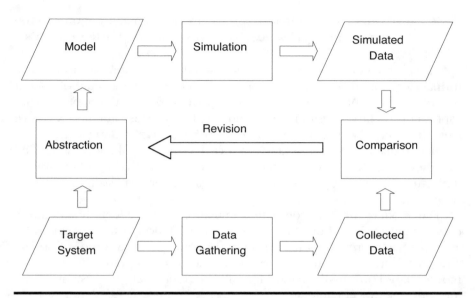

Figure 4.1 Simulation Model Definition and Validation Scheme

Different classes of models are characterized by different elements used to represent the real system or phenomenon, as well as by different rules coded in the simulation program to execute the model. Thus, two fundamental issues are (1) the choice of the most appropriate class of models; and (2) the definition of a model for the target system that can be trusted. The latter issue is generally referred to as the model validation problem. For example, if a decision-maker (DM) wants insight from a simulation about how to select the best inventory management policy from among a set of alternatives in order to reduce the storage and supply costs of a manufacturing company, the DM must be quite sure that the outputs from the simulation model of the context that he or she wishes to study do not significantly differ from the actual ones.

The general scheme of the simulation model definition and validation is illustrated in Figure 4.1, which highlights the following phases:

- Abstraction: a model of the target system to be analyzed is defined by an abstraction process.
- Simulation: the behavior of the model is observed by executing simulation runs that generate a collection of simulation data.
- Data gathering: data describing the actual behavior of the target system are collected.
- Comparison: the simulated and real data are compared and the level of appropriateness of the model is evaluated.

■ Revision: possible variations in the model to make it closer to the target system are introduced by revising the abstraction process.

The abstraction process generates a simulation model consisting of a formal description of the system or process. The level of detail of such a description must be suitably defined so that only the subset of relevant aspects of the target system is reproduced. This process implies extensive simplification of the real-world system to be analyzed and, as a consequence, enhances the importance of a correct model validation. Provided that they are able to show a valid behavior, i.e., one that is coherent with the actual evolution of the system or process under analysis, models must be kept as simple as possible.

The models used in computer simulation share a common feature: they are always based on some formal representation of the system under analysis. Such formalizations, in fact, must correspond to some set of data and procedures that can be coded in a simulation program. For example, models based on a mathematical formalization represent the system behavior with a set of equations.

Conventional simulation models fall into one of several categories. *Dynamic models* are able to reproduce the evolution of a system over time, whereas *static models* assume the system operates in a certain fixed time instant (for or example, the well-known Monte Carlo simulation). If a simulation model does not include any random variable, that is, it always provides the same output for a fixed set of input data, the model is deterministic. *Deterministic models* can be exploited in manufacturing, for example, to implement scheduling heuristics [3].

Conversely, *stochastic models* include random variables to reproduce the possible occurrence of events or disturbances that are unknown *a priori*; such models need a formal representation of stochastic phenomena, which is usually given by a set of probability distributions and a set of relevant statistical parameters (such as means and standard deviations) to generate suitable values for the random variables. The analysis of the system's behavior performed through a stochastic simulation model corresponds to a statistic experiment, which must be correctly planned and whose results must be carefully interpreted by means of statistical tests.

Simulation models can be *discrete*, *continuous*, or *mixed*. Discrete models change their state only at discrete instants with time, that is, the system's state can be described by a number of variables whose values are only updated at such discrete instants. Discrete simulation models are based on the assumption that nothing relevant happens between two successive state changes during the model simulation, and that this does not introduce a simplification that invalidates the model's behavior compared to that of the corresponding actual system. Continuous models, on the other hand, are often described by a set of mathematical equations, such as ordinary or partial differential equations, and change their state continuously over time. Finally, mixed models include discrete and continuous state variables. Note that in computer simulation, any possible computation process is intrinsically discrete; therefore, even for continuous models, the system's evolution is generated in correspondence to a sequence of discrete time instants (e.g., synchronously with a prefixed reference time interval).

As long as a suitable model has been defined for a system, the simulation can be performed following different methods. Whenever a system model consists of a set of equations, the simulation of the system's evolution can be performed in an analytic or a numeric way. The former corresponds to computers used to solve the equations, e.g., through differential calculus, whereas the latter entails the use of a computational procedure to execute the model in order to generate an artificial history of the system's evolution that can be (statistically) analyzed. As an example, in some simple cases queuing theory problems can be solved by the analytical study of policies used to dispatch incoming operation requests to the machines available on a shop floor; however, queuing models most often reveal their properties only through a numerical stochastic simulation.

Discrete-event simulation (DES) is the branch of simulation methodology widely adopted for the analysis of manufacturing systems. DES is based on discrete stochastic models. As a matter of fact, the simplification introduced by assuming manufacturing processes as discrete ones is usually accepted. This means, for example, modeling the way a welding machine performs an operation on a material with time; what is important, rather, is to know at which time instants the machine state changes, i.e., when the welding operations start (thus changing the machine state to busy) and when they finish (when the machine state returns to idle).

Several alternative methods exist to define DES models, because they can be specified following different conceptual views of reality; each provides model designers with different representation schemas. The basic approaches to formalize a system with DES correspond to the event-scheduling and process-interaction views. The choice of one of the approaches over the others influences the conceptual definition of the simulation model and, in particular, specifies the algorithm needed for its execution; in fact, event scheduling and process interaction deal with the dynamic structure of a DES model.

With event scheduling, the focus lies on events whose occurrence may involve one or more entities in the system and whose effects are implemented by event processing routines responsible for updating the state of the system. A particular type of event-scheduling simulation paradigm is the so-called *activity-scanning* approach, which distinguishes two classes of events: determined events, which depend on current activities (e.g., the duration of the machining of a part) or on exogenous occurrences (e.g., the arrival of a new job), and the contingent events, which occur when some conditions in the system state are verified.

In the process-interaction approach, a simulation model consists of a set of processes that correspond to a sequence of events and activities. For each process, a procedure should be defined that specifies how the process evolves over time and under which conditions the process execution must be suspended and restarted. In order to identify the relevant processes, the entities present in the system must be taken into account. Therefore, the processes model the sequence of events and activities that can influence the entities' existence in the system, and the whole system's evolution emerges from the interaction among such processes. For the event-scheduling approach, the execution of a simulation is based on a main procedure that manages the *future event list* and invokes the appropriate event-processing routines; a simple general algorithm used to run DES based on the event-scheduling approach is shown in Figure 4.2. By contrast, in the case of the process-interaction approach, the simulation consists of the concurrent execution of the threads relevant to the set of "active" entities in the system.

It must be observed that event-scheduling and process-interaction modeling paradigms tend to define a system as a whole. As a matter of fact, even when system entities to identify the relevant processes are considered, the variables representing the system state are typically stored as global ones managed by means of a shared memory. The *logical process* concept [4] was introduced to extend the preceding conceptual views in order to allow modularity in system modeling. Following such a paradigm, in fact, the attention of human modelers is actually geared toward the entities composing the system, so that the system state is partitioned into subsets of state variables associated with those entities, and no global shared state is used. With the logical process view, a simulation corresponds to a set of logical processes involving the system entities, which evolve by exchanging messages and modifying only a local subset of state variables.

It should be apparent why the natural implementation of tools and languages for DES based on process-interaction and logical processes has also led to the exploitation of the object-oriented design paradigm for simulation: using an object-oriented design, the relevant system entities are modeled as objects defined by a set of private data describing the entity state and methods implementing the processes ruling the entity evolution. Most of the simulation packages or languages currently adopted in manufacturing are object oriented or allow an object-oriented representation of the systems. A review of such tools is beyond the scope of

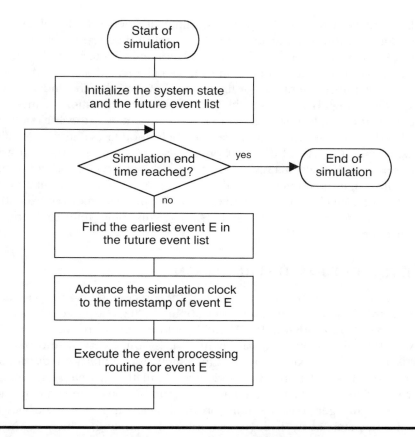

Figure 4.2   General Algorithm for Event-Scheduling DES

this chapter and deserves further attention; for additional information, refer to a recent survey by Swain [5].

The previous chapters placed in evidence the tendency of modern manufacturing systems toward decentralization. Logical process-based simulation, implemented by means of object-oriented simulation languages or tools, actually allows the development of component-oriented models to study the behavior of distributed systems. Distributed DES programs are composed of a set of logical processes that can communicate with each other through a shared memory or by exchanging messages. In distributed DES, the purpose of logical processes is to simulate the components of a distributed system (a quite recent reference about this field is Fujimoto [6]).

Distributed DES models are clearly suited for a distributed parallel execution; an advantage of this possibility lies in the scalability of simulation models that become able to deal with large complex systems as multisite manufacturing enterprises or enterprise networks are. However, in parallel distributed simulation the treatment of time represents a critical issue [7]. In contrast to centralized simulation, distributed simulation has no global control of the simulation time clock; each process manages its own list of future events according to its local clock. The independent execution of logical processes therefore risks the generation of inconsistencies in the whole system's evolution; for instance, an event could be simulated for a given process in real time before other events with smaller timestamps in the lists of different processes. Research on parallel distributed simulation deals mainly with the development of synchronization algorithms that guarantee the correct evolution of the simulation by introducing a small computation overhead (i.e., more memory or execution time requirements). For further details about this topic, refer to Tropper's recent overview [8].

MULTIAGENT-BASED SIMULATION

Some of the observations at the end of the preceding section provide a good starting point for a discussion about multiagent-based simulation. As observed by Davidsson [9], "MABS should not be seen as a completely new and original simulation paradigm ... it is influenced by and partially builds upon some existing paradigms, such as parallel and distributed discrete event simulation, object-oriented simulation, as well as dynamic micro simulation." In fact, as long as agents can be considered an extension of objects, an agent-based simulation model can in principle be viewed as a further step away from object-oriented modeling, which defines a system as a collection of active and autonomous entities, i.e., agents. This could very well be the viewpoint of software engineers enthusiastic about the new opportunity of exploiting the agent-based modeling paradigm as an even more flexible tool to design effective applications.

As we will remark later, MABS has been judged suitable to deal with discrete and distributed systems, so its link with parallel and distributed DES should be clear. Dynamic microsimulation aims at studying systems that focus on the evolution of the single individuals (usually a sample extracted from a population) that make them up. This technique, however, which is applied, for example, to analyze economic, biological, or traffic systems, does not explicitly consider the interactions among individuals. MASs, and likewise MABSs, are founded precisely on the modeling of individuals as agents.

At this point of the discussion the differences between the previously mentioned simulation paradigms and MABSs could be deduced even from what should be known about agents and MASs. Nevertheless, the fundamental aspects characterizing MABS models will be outlined in this section. It must be borne in mind that we are not advocating MABSs as the solution to every simulation requirement; on the contrary, we would tend to proceed quite cautiously. Indeed, the questions to be answered concern what MABSs means and their usefulness for improving manufacturing performance.

As noted by Hare and Deadman in a contemporary analysis [10], many different terms in the literature have been used as synonyms of ABS, namely, "agent-based simulation modeling; multiagent simulation; multiagent-based simulation (MABS); agent-based social simulation (ABSS); [and] individual-based configuration modeling," as well as MASs. In an attempt to sort out this alphabet soup of terminology, the authors underlined two key peculiar features of ABS: the capability of model interaction among individuals and the fact that such interactions derive from the use of deliberative knowledge by agents.

The term used here, "MABS," can be defined as the modeling and simulation of real systems consisting of intelligent agents that cooperate with each other; thus, the simulation is specified as multiagent based because the simulation model includes many interacting agents. According to the AgentLink Roadmap [11], "agent-based simulation is characterized by the intersection of three scientific fields, namely, agent-based computing, the social sciences and computer simulation." Typical application areas for ABS include the simulation of economic, societal, and biological environments.

The specific differences between MABSs and conventional simulation models are summarized by the following points:

- Part of the system entities is associated with agents.
- The entities that are modeled with agents can communicate with one another, perceive changes in the environment, and show a proactive behavior.
- The system model is intrinsically distributed because agents behave autonomously.
- A new flexibility is allowed for the system evolution because agents can be created and destroyed dynamically.
- MABS models make it possible to study the emergent behavior of a system, i.e., the outcome of the simulation at the macro level derives from the evolution of the interaction of single or groups of agents at the micro level.

Agent-Based Simulation in Social Sciences

The possibility of analyzing the emergent behavior of a complex system highlights the close relationship between MABSs and the empirical study

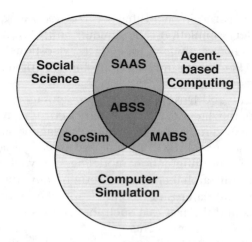

Figure 4.3 Map of the Scientific Fields Derived from the Intersections of Computer Simulation, Social Science, and Agent-Based Computing (Redrawn from Davidsson, P., J. Artif. Soc. Social Simulation, 5, 2002, available at http://jasss.soc.surrey.ac.uk/5/1/7.html)

of social science models. Davidsson [12] characterizes agent-based social simulation (ABSS) as the field lying at the intersection (shown in Figure 4.3) of the three scientific areas mentioned previously, namely:

■ Agent-based computing, corresponding to the research area of computer science devoted to agent-based modeling, design, and programming.
■ Social science, the research field that studies the interaction among social entities and includes social psychology, management, policy, and some areas of biology.
■ Computer simulation, which deals with the study of techniques to simulate phenomena and systems with computers.

For a complete understanding of the map depicted in Figure 4.3, one can consider the additional fields associated with the three partial intersections of the preceding areas. The social aspect of agent system (SAAS) deals with the study of norms, institutions, organizations, competitions, etc. The social simulation (SocSim) focuses on the possibility of simulating social phenomena with computers using any available simulation method and model. The likening of MABSs to an intersection between agent-based computing and computer simulation can be accepted considering that we are interested in analyzing the use of ABS in connection with manufacturing systems and processes. However, we believe that social science also affects MABSs to a certain extent because, for example, some social interaction models could be required to simulate MABS phenomena inside a manufacturing system or along the supply chain. As a matter of fact, although we are specifically interested in manufacturing, we believe that understanding the use made of ABS in social sciences helps to recognize what ABS can do, and therefore what we can expect from it, even in other applications.

Numerous works have underlined the importance of simulation as a tool to study social and biological systems, and more recent studies have vaunted the effectiveness of ABS for the analysis of these fields in comparison to "traditional" simulation approaches (for example, consider Troitzsch [13]). Conventional analysis methods represent the social system behavior through a set of mathematical relationships, e.g., differential equations. The introduction of computers in the study of social and biological phenomena exploited the computer's ability to manipulate mathematical symbols automatically, thus allowing the use of equation-based models.

The advent of agent-based modeling represented a remarkable inno-vation; with this type of modeling, the behavior of complex social systems, made up of multiple active entities, could be simulated by modeling and simulating the single components and their interactions (modularity). In addition, the system behavior did not need to be embedded *a priori* into a set of equations, but could emerge from the whole of the behaviors of the single entities interacting with each other and with the environment (emergent behavior). This latter feature obviated the *a priori* introduction of strong assumptions, such as the rationality assumption of decision-makers necessary for game theory-based models.

In agent-based models, individual behaviors can adapt to circumstances as well as self-organize, showing a sort of learning ability. ABSS permits the analysis of the emergent behavior characterizing a population at an aggregate level (for instance, discovering the attitude of a population to favor the reproduction of its best individuals to ensure the survival of the species). ABS models in the social sciences typically include a large number of autonomous entities, which interact and evolve following simple rules. Therefore, knowledge and decision responsibility are highly distributed in such systems and, consequently, the observed emergent behavior is usually unpredictable *a priori*.

When Can MABS Be Applied? Some Considerations and Conclusions

What should be clear at this point is that the focus of MABS modeling is on behaviors; the agents are entities with some defined behavior, and the interaction among such entities may originate in a complex system evolution. Thus, a first conceptual conclusion about the appropriateness of MABSs to analyze the dynamics of a system should stem from whether that system can be suitably modeled in terms of entities and behaviors. For example, when a manufacturing subsystem is characterized by a batch or continuous production process whose representation (in terms of flows and levels and given by a set of equations) is considered a valid model by the company decision-makers, no apparent reason exists to force an agent-based solution. Actually, in such cases, global system state variables are suitably computed from equations, and no hidden behavior can emerge from individuals because nondeterministic outcomes can be correctly modeled as random disturbances. Let us, on the other hand, try to identify the situations inside or outside a manufacturing company that can actually benefit from MABSs.

Presence of Autonomous Entities

The MABS is the right tool to analyze situations in which distributed entities with an autonomous behavior are present. This explains the increasing number of MABS applications for the analysis of the business process along the manufacturing supply chain (consider Moyaux et al. [14], Szirbik et al. [15], and Amin and Ballard [16] among the most recent examples). In a supply chain or network, several distinct entities must interact in a coordinated way to reach specific and shared objectives, e.g., to satisfy customers' requests with quality products within the due dates, while reducing the product cost with an effective integrated production management policy. Entities correspond to production and distribution enterprises, linked by supplier–consumer relationships, and final customers. In the supply network, single entity behavior is clearly influenced by those of other entities; as a consequence, the dynamics of supplies, inventories, and their relevant costs can be thought of as a complex system behavior emerging from the single actors' decisions and processes.

Parunak et al. [17] discussed the suitability of ABS in the context of the Dynamic Analysis of Supply Chains (DASCh) project, comparing in particular agent-based to equation-based modeling capabilities in handling the well-known *bullwhip effect* in supply chains. These authors offer an interesting general recommendation: agent-based modeling is appropriate for domains characterized by discrete decisions and composed of a high number of distributed local decision-makers; on the other hand, equation-based modeling better suits centralized systems whose dynamics are characterized by physical laws rather than information processing.

Another application motivating the presence of autonomous entities, which will not be discussed in this chapter, is the use of MABSs with very frequent interactions with the human user, thus simulating a complex system and working as educational agents to train personnel to interact with that system.

Analysis of Distributed Architectures and Control Policies

The previous section underscores the correctness of using a multiagent-based model to simulate contexts presenting autonomous active components that play an agent role (we use the term *agent* just because many things act like agents). Here, we will ease up on the autonomy requirement, focusing instead on the architecture of the system under analysis. We claim, in fact, that the MABS is highly appropriate for studying the evolution of distributed systems and, in particular, for evaluating the performance of decentralized decision policies. Such situations occur increasingly more frequently in manufacturing systems, where distributed decision-making has been labeled as a fundamental building block enabling agility. In this context, two aspects relevant to the study of strategies for decentralized control (here the term *control* generically denotes any of the planning, scheduling, and control activities in manufacturing) are particularly interesting:

- The analysis of individual control strategies (agent behavior), in which the purpose is to develop individual strategies. Manufacturing entities (e.g., machines, tools, storage areas, and so on) associated with agents can follow these strategies in order to achieve a common goal, without imposing a complex central regulation system
- The analysis of coordination strategies (emergent behavior), in which the interaction among the single entities must be ruled by a coordination mechanism that, for example, establishes how the agents cooperate and resolve possible conflicts. Clearly, the performance yield of the system as a whole depends heavily on the kind of coordination adopted; thus, an objective of MABSs could be to make such dependency clear — in other words, to support the definition of an effective coordination strategy.

The relevance of using MABSs in evaluating the performance of complex technical systems that are distributed and involve interaction with humans is also emphasized by Davidsson [12]. Note that no requirement has been imposed about the possible implementation of the distributed control architecture; this could not be agent based, but could nevertheless be simulated with an agent-based simulator. It is evident that in the case of an agent-based control system, the choice of MABSs is even more appropriate if not mandatory.

We can thus conclude that the MABS is an important simulation technique in the current manufacturing scenario because the agility requirement is strictly connected with the increasing need for highly decentralized systems inside companies and with the need for continuous coordination among the entities along the supply chain. The reality that manufacturing enterprises are currently living can be likened to a web of complex relationships among entities that, at the different levels and scopes, act more as peers than as hierarchical components. The possible role of the MABS here is that of forecasting the performance levels emerging from such relationships and, at the same time, supporting the design of effective individual and coordination strategies.

Modeling and Implementation Issues for MABSs

Having discussed the applicability of MABSs in manufacturing, let us now turn to some technical aspects about its implementation. More specifically, we want to establish what the architecture of an MABS is like, and what is needed to execute an MABS model correctly, that is, to run the agent-based simulation. An MABS model basically consists of two main parts:

■ The MAS model specifies the active entities of the system as agents, each characterized by its own knowledge and behavior. Entities, and therefore the associated agents, can correspond to physical and logical elements and can be permanent or transient. The subsets of system state variables relevant to active entities are also included in the agents' private data, so that only the agents have the responsibility for storing and updating them. The agents are active because they are capable of perceptions, communications, and actions; their methods finally control the system's processes and the state's evolution. In addition, the MAS model defines the ontology (the messages that agents can exchange and understand) and multiagent architecture (e.g., hierarchy, federation or autonomous agent-based) are used.
■ The model of the environment includes (1) the passive entities in the system (i.e., not capable of an autonomous behavior) and their relevant state variables; and (2) any possible elements situated in the physical world (such as suppliers, customers, or competitors) and in the information world (e.g., databases, ERP systems, MRP modules) that exogenously influence or are necessary for the system's evolution, but whose state is not part of the "controlled" system. In an object-oriented view, while agents are used to model active entities, objects can represent the passive ones contained in the environment.

In an MABS model, the system state is thus distributed and mainly controlled by separate agents. Their methods implement the processes involving the entities and so they characterize the system's evolution. The agents' behaviors, the ontology, and the MAS architecture together define the way single entities or groups of entities interact. The environment model also denotes the relationships among system entities and anything else needed to simulate the influence of the world surrounding the system, such as user interfaces, exogenous inputs, disturbances, or stochastic variations.

In order to execute an MABS model, a simulator engine is needed — that is, a program able to feed the model with the exogenous (possibly) random inputs and to manage the time and the system's state. In an MABS, the evolution of the state of the entities (agents and objects) is discrete; MABSs can thus be simulated with DES, more precisely with MABDES (multiagent-based discrete-event simulation). In addition, having distributed the system state to agents and objects, the execution of an MABDES can be implemented as a parallel DES, where multiple threads, activated by the agents and by the routine that manages the environment, can run concurrently. As already pointed out, such a possibility greatly enhances the modeling and computational capabilities of the simulation, but implies the need of dealing with the problem of time synchronization. However, for the sake of simplicity, the issues relevant to parallel and distributed simulation are not considered here.

To illustrate the possible basic steps of an MABDES cycle briefly, we consider the abstract architecture and the execution model proposed by Wagner and Tulba in a recent work [18] in which the authors devise an agent-oriented modeling approach based on the enrichment of the UML formalism to include the concepts of the agent–object-relationship (AOR) metamodel [19]. The kind of simulation of Wagner and Tulba [18] is a time-driven discrete-event one, because the simulated time advances in small (unitary) regular time steps. In this way, the perception–reaction cycle of agents can be easily simulated because the reaction to a perception occurring at a time t follows in the immediately successive time $t + 1$.

The simulation model consists of active objects, i.e., agents, and passive ones; in addition, the state of agents is separated into an internal state, which is strictly relevant to the agent reasoning activities, and an external state, which is relevant to the physical entities with which the agents are associated. The state of a simulated system thus consists of the simulated time; the environment state (i.e., object and external agents' state); the internal agents' state; and, finally, the list of future events. The simulator engine of Wagner and Tulba [18] has two components:

■ The environment simulator manages the state of all passive objects and the external state of each agent and is responsible for managing the influence of the future events, ordered in a future event list, on this portion of the system state.

■ The agent simulator is replicated for each active agent and manages the agent activity by updating the internal state of agents according to the occurrence of the relevant perception events, i.e., the events used to simulate the agent perception of changes in the environment and communications.

A simulation cycle used by Wagner and Tulba [18] consists of the following steps:

1. The events that must occur at the start of a new simulation cycle (e.g., at a simulated time *t*) are identified by the environment simulator in the future event list, including also the exogenous events, i.e., stochastic events or events created by actors that are external to the system. If the future event list is empty, the simulation ends.
2. On the basis of the current environment state and the current events, the environment simulator determines the new environment state; the set of new events that must be inserted in the future event list as a consequence of the processing of the current events; and the perception events for each agent.
3. For each agent, the associated agent simulator, on the basis of the current internal state and the current perceptions, computes a new internal state and a set of events relevant to the consequent actions performed by the agents to be added to the future event list with a timestamp $t + 1$.
4. The future event list is updated by removing all the processed events and adding the new computed ones.
5. The environment simulator updates the simulated time *t* by incrementing it by one, then starting a new simulation cycle.

We believe that the architecture for an MABS proposed by Wagner and Tulba [18], even if suited for a single processor execution, can be considered a valuable general reference. In fact, distinguishing the simulation model into one external and more internal state components allows the model to represent with modularity the environment and the active actors present in the specific situation to be simulated. Interestingly, these authors [18] point out how the combined use of an internal perspective (i.e., one that views the system from the single agent standpoint) and an external perspective (i.e., one that observes the system from the outside as a whole) can be used to characterize the methodology for developing an AOR simulation model.

PLATFORMS TO SUPPORT MULTIAGENT-BASED SIMULATION DEVELOPMENT

The literature is replete with reports devoted to the development of MABS platforms to support the design and implementation of agent society simulations. As a matter of fact, several frameworks exist but, for the sake of brevity, this section will not seek to provide a comprehensive survey of them. Because of the interest they may arouse, however, it will comment briefly on the evolution that these environments have had throughout MABSs' brief history. According to a recent review by Gilbert and Bankes [20], platforms and methods for agent-based modeling have evolved from conventional programming packages to the distribution of libraries of routines to achieve the dimension of packages allowing a simplified design of an agent society through the provision of some visual interface.

Subsequent to the first attempts to implement an MABS in C++, the first MABS libraries, Java and SmallTalk (developed mostly in Java language), appeared in the early 1990s. Also emerging were Swarm [21] and RePast [22], two of the most often quoted libraries in the literature, which are briefly described in the following subsections. The evolution to packages supporting MABS development is justified by the aim to allow users inexpert in programming to develop their own models. The first packages thus offered easy environments (such as StarLogo [23] and AgentSheets [24]) that enabled the design of very simple models with some limitations in functionality [20]. To overcome this drawback, subsequent packages have become increasingly more complex and powerful to the extent that, although it is not always necessary to be a programmer to use them, the time and effort needed to learn them has risen in parallel [20]. Examples are SDML [25] and Desire [26]. An additional trend has been to specialize these packages for applications in specific domains, as is the case of Cormas [27], with a specificity in the domain of natural resources management, and of MAST [28], specialized in manufacturing. MAST is briefly described in the third subsection that follows.

Swarm

In 1994 a team of researchers led by Chris Langton started the Swarm Project [21] in order to create a standard support tool that could manage "swarms" of objects needed to develop MABS models. Swarm is an open and extensible framework released with a GNU-GPL license, developed in Objective C and more recently in Java. The first version dates to 1995 and the second to 1997; both were based on UNIX, Solaris, and Linux operating systems and on Objective C language. The subsequent creation of a Java layer that shows the Swarm library as Java interfaces has allowed further spread and uptake in other computer environments.

The core of Swarm is an object-oriented framework that defines the behavior of agents and of the other objects' interacting during a simulation. The definition of an agent is related to the definition of its behavior, that is, its action rules. An agent is a container of rules answering to specific stimuli; in addition, it also contain a swarm of other agents that allow the processing of nested structures to represent very complex realities. When agents are defined with their own specific characteristic, it is possible to specify MABS models by creating a network of links among agents as well as between agents and the environment, thereby enabling the modeler to define agent behaviors and interagent transactions. The Swarm schedule library allows creating a simulation clock. Swarm also facilitates interaction between the user and the model through the use of probes, which allow the insertion of key variables directly from a dialogue interface on the screen. For example, the probes can improve management of charts and can be used to modify the simulation input values from the windows on the screen.

Work by Strader et al. [29] offers an example related to manufacturing in which a model developed in the Swarm multiagent simulation platform is used to study the impact of information sharing on order fulfillment in divergent assembly supply chains (commonly associated with the computer and electronics industries).

RePast

RePast [22] is the acronym for recursive porous agent simulation toolkit and is a framework developed by the Social Science Research Computing Center of the University of Chicago to create agent simulations using Java language. At the outset, RePast was viewed as a set of libraries intended to simplify the use of Swarm. It was then redesigned completely in Java as a new framework, making use of some abstraction keys of Swarm. A recent work by Tobias and Hofman [30] proposing an evaluation of free Java libraries for social–scientific agent-based simulation judged it to be the most suitable simulation framework for the applied modeling of social interventions based on theories and data.

RePast provides a library of classes to create, perform, view, and collect data from agent simulations. Moreover, RePast includes different charts to view data, creating photographs of the objects on the screen and Quick-Time videos of a running simulation. It can be defined as a "Swarm-like" tool of simulation because it has the main characteristics of Swarm, but also includes a run time manipulation model controllable by graphical interfaces. In RePast, the simulations are similar to a state machine, whose states are created by the global states of all its components that are divided into *infrastructure* and *representation*. The infrastructure performs the simulations and views and collects data; the representation is what the modeler makes — in other words, the simulation model. The infrastructure state is the viewing state, the state of the object that collects data, etc. The representation state is the state of what has been modeled, the current values of all the agent variables and of the space in which they operate, and of all the represented objects.

In RePast, as in Swarm, every change of the infrastructure components and of the representation components happens through a schedule. Put succinctly, RePast lets users build a simulation in which all the changes pass through a schedule, as occurs in a DES. Models developed with RePast present simple scheduling, as well as more dynamic and sophisticated mechanisms like the execution of an event that schedules itself for future events.

Further reading can be found in Größler et al. [31], who provide an example related specifically to supply chain management in manufacturing built on a software-based integration of RePast agent-based modeling and system dynamics simulations.

MAST

An example of an MABS approach devoted entirely to the manufacturing world is the MAST [28], manufacturing agent simulation tool, developed at the Rockwell Automation Research Center in Prague. Programmed in Java and built on top of the JADE agent platform [32], the MAST is designed mainly for the simulations of material handling systems. It provides the user with the agents for basic material handling components, e.g., the manufacturing cell; conveyor belt; diverter; automated guided vehicles (AGVs), etc., so that various material handling system configurations can be modeled. In the simulation phase, the agents transport discrete entities among manufacturing cells and cooperate together via message sending using a common knowledge ontology developed for the material-handling domain.

Tobias and Hofman [30] excluded *a priori* JADE middleware from their evaluation because, in their view, JADE can only provide assistant agents and mobile agents and is not really suited for theory, for social scientific, or for data-based simulations. Nevertheless, for the reasons that follow, the MAST represents a significant step forward as regards holonic manufacturing modeling and significant simulations of agility.

In the first place, this simulation tool is made for manufacturing, and it has been demonstrated on three main manufacturing tasks: conveyor-based transportation, AGV-based transportation, and assembly tasks. As a consequence, a strong emphasis is devoted to failure detection and recovery and to dynamic reconfiguration issues, which are historically among the main important results requested of simulation in manufacturing. Other characteristics that make the MAST a very promising tool for manufacturing include its compliance with the FIPA standard (due to the adoption of JADE); its precise collocation with respect to the function blocks given by the IEC-61499 standard (see Chapter 3); and its use of XML as a general language for message content and for the description of the world in which agents act.

Another interesting feature of the MAST can be interpreted to some extent as in favor of as well as against its use: due to its characteristics and its heavily JADE-dependent implementation, the separation of the MAST from its real aim (that is, as a simulation tool based on agents specific for manufacturing) and its possible use as a methodology for MAS design implementation, is not so wide. As a matter of fact, the graphic user interface of MAST already allows the design of user-defined MASs. In addition, as assessed in the conclusions of Vrba [28], the developed agents, written in JAVA language, will soon be run on standard PLC-based automation controllers in parallel with low-level, real-time control code (ladder logic).

AN APPLICATION OF MABS IN PS-BIKES

The purpose of this section is to briefly illustrate a possible application of an MABS as a decision tool to analyze some alternatives for the introduction of MASs in PS-Bikes. In particular, the MAS devoted to the scheduling activity in the framework area introduced in the section in Chapter 3 on an MAS model for planning and scheduling in the new PS-Bikes plant is considered. According to the MAS model proposed there, the schedule emerges from the negotiation between the job agents (JAs) and the machine agents (MAs), whose behavior is ruled by several selection functions: the request probability selection function (RPSF) and the bid selection function (BSF) of JAs and the acceptance probability selection function (APSF) for the MAs. In turn, the structure of such functions depends on the values given to a set of parameters that have so far been assumed to be fixed by PS-Bikes' scheduling administrators on the basis of their experience or some heuristic reasoning.

Obviously, because the choice of the values for the MAS scheduler parameters could be critical, it seems appropriate to perform a tuning process in order to identify suitable values for them. This entails designing a system that allows the MAS scheduler, through a sort of learning capability, to adapt itself to the typical problem instances occurring in the PS-Bikes' framework area. To this end, a possible approach is that of considering the scheduling cost, yielded by the MAS scheduler, as the output of a cost function, $f(\theta)$, whose structure is unknown and can thus generally be assumed to be a nonlinear one, and which takes the vector of the scheduler parameter values, θ, as inputs. The parameter tuning system can then be devised as a nonlinear optimization procedure that systematically changes the values of the input parameters in θ, seeking to improve the output performance returned by $f(\theta)$.

The architecture of a tuning system built along these lines is drawn in Figure 4.4 and will be discussed in the next subsection. Nevertheless, the figure hints at the role played by the MABS in this scenario: the MABS allows reproducing the behavior of the MAS scheduler in order to view and evaluate the influence of the parameters' different values on the scheduling objective rapidly, that is, to compute the value of the cost function $f(\theta)$ properly. In addition, as will be described later, the MABS model introduced in this section also allows verifying the influence of new machines (i.e., laser cutters) introduced in the PS-Bikes framework area on the scheduling performance.

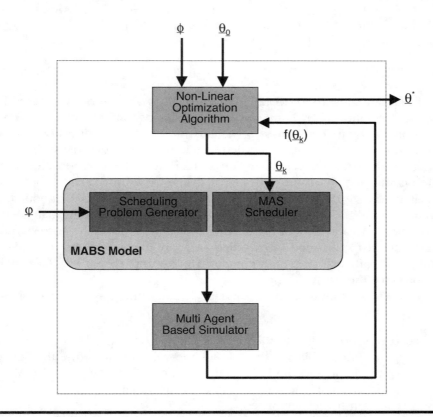

Figure 4.4   Architecture of the Tuning System for the Parameters of the MAS Scheduler

MABS System for Tuning MAS Scheduling Parameters

Some further details about the MABS-based tuning system whose architecture corresponds to the scheme in Figure 4.4 are useful. Two boxes in the figure denote the two main procedure modules, i.e., the nonlinear optimization algorithm (NLOA) and the multiagent-based simulator (MABSim); the rounded-corner box indicates the MABS model. The core procedure that makes the tuning system work is the NLOA, which iterates by generating new configurations of the parameters, taking into account the configurations previously used and the value obtained for the objective cost function in order to minimize the latter. At each iteration, the NLOA generates an MABS model with a modified parameter configuration; this model is then executed by the MABSim, which finally feeds back the computed schedule cost to the NLOA.

Let us now consider the rationale for this tuning system. As already stated, the MAS scheduler can be thought of as a generic nonlinear function, $f(\theta)$, of the scheduling parameters, θ, whose structure is unknown. From this assumption, the general structure of the NLOA adopted is given by the following recursive equation:

$$\theta_{k+1} = \theta_k + t_k \cdot d_k \qquad (4.1)$$

This computes a new configuration for the parameters to be used at the next iteration, $k + 1$, on the basis of the parameters at the current iteration, k, of an iteration step, t_k, and of an iteration direction, d_k — both chosen so that $f(\theta_{k+1}) \leq f(\theta_k)$, i.e., the value returned by the cost function, does not increase. Nonlinear algorithms of this kind differ from one another mainly in the way in which the iteration direction (but also the iteration step) is computed [33]. Because no information about the structure of $f(\theta)$ is known in this case, an algorithm is used that needs only the information about the sequence of θ_k tested and the corresponding values $f(\theta_k)$ obtained (a so-called direct nonlinear optimization algorithm); in particular, the algorithm by Hooke and Jeeves [34] has been adopted for the analysis under concern.

Three sets of input data are needed by the tuning system (as shown in Figure 4.4):

- θ_0, corresponding to the initial configuration of the MAS scheduling parameters. This vector includes the values that define the shape of the RPSF and the BSF for the JAs and of the APSF for the MAs, and represents the starting values of the variables affecting the cost function to be minimized;
- φ, including the values for the parameters describing the statistic properties of the scheduling problems to be solved by the simulated

MAS scheduler (e.g., the job interarrival time, the distribution of job processing time, etc.); the number of available machines; and the maximum number of jobs to generate at each optimization cycle (that determines the duration of the simulation). This information is used by the scheduling problem generator (SPG) contained in the simulation model to create the problem instances randomly; in addition, the values of the seeds for the random number generators used by JA, MAand SPG are specified in this vector.

■ ϕ, containing the values of the parameters needed by the NLOA. As an example, the parameters needed to determine the stopping conditions for the optimization cycle must be specified.

The MABS-based parameter tuning procedure corresponds to the iterations of the optimization cycle performed by the NLOA. In particular, at each iteration k:

1. The NLOA invokes the simulation of the MAS scheduler of the MABS model by setting the values for the parameters as in θ_k.
2. During the simulation, the SPG included in the MABS model randomly generates a sequence of job arrivals that are communicated to the simulated MAS scheduler.
3. The simulated MAS scheduler schedules the jobs over time by following the negotiation protocol among JAs and MAs described in the section in Chapter 3 concerning an MAS model for planning and scheduling in the new PS-Bikes plant.
4. At the end of the iteration, whenever all the arrived jobs have been processed, the MAS scheduler simulated by the MABSim returns the value of the cost function, $f(\theta_k)$, to the NLOA.
5. The NLOA verifies whether the stop condition has been satisfied, and in this case it terminates the optimization cycle by outputting the vector θ^* containing the best parameter values determined; otherwise, it starts iteration $k + 1$ from step 1.

The role of the MABS in this setting is thus that of simulating the activity of an actual MAS, the scheduler for the framework area presented in Chapter 3, which allows computation of the scheduling cost function in correspondence of any given configuration of the parameters affecting the agents' behaviors.

To provide some helpful advice to PS-Bikes' managers, a prototype of the tuning system, and thus of the MABS, was developed in Java code. Next, a few key points about this simulation software will be discussed, referring to the architecture of the MAS scheduler and the relevant agent negotiation protocol proposed in Chapter 3.

The parameter tuning systems and the MABS are based on the class diagram depicted in Figure 4.5. The class *Algorithm* represents the core of the system as it implements the optimization algorithm; a method of this class (f(double[], int)) computes the scheduling cost function for the given configuration of the parameter vector at each iteration. Such a method invokes the execution of a simulation run, thereby activating the relevant methods of the class *Model*. As can be observed in Figure 4.5, all the class components that implement the MAS to be simulated depend on the class *Model*.

As already pointed out, our interest here goes beyond the characteristics of the agents in the MAS model, because these basically replicate the ones discussed in Chapter 3. However, we would like to emphasize how the simulation cycle is performed, that is, to discuss how the MABS deals with time evolution.

The MABS proceeds as a time-driven discrete-event simulation; in fact, a method called step() of the class *Model* is used to generate the sequence of time instants t, at whose occurrences the activities of the active agents are triggered and synchronized. The step() method represents the actual simulation engine because it is responsible for controlling the system state evolution and the generation of the exogenous input from the environment. The step() method:

■ Triggers the possible new job arrivals at time t.

■ Updates the state of the machines associated with the MA; in case the processing of a job on a machine is completed at time t, the machine becomes idle. On the other hand, if a job must start its processing on a machine according to the associated MA agenda (the private local schedule), the machine becomes busy.

■ Verifies the state of the JAs ready for scheduling and activates the JA process that (according to what is illustrated in Chapter 3) determines whether the JA will ask for service to the MAs and, if the answer is yes, it adds the JA to the list of JAs that, at the next time $t + 1$, will start a negotiation by sending a message to the MAs.

■ Activates the JAs ready to start a negotiation, so that they compute their bid and send a request for service message by invoking the methods of the *MessageDispatcher* class.

■ Activates the MA process that checks the arrivals of new requests from the JAs, that decides whether to serve the JAs and which time slot to offer or not, and, if necessary, that sends the appropriate reply message to the JAs.

■ Starts the service offer evaluation for the JAs that at time $t - 1$ received an MA proposal, allowing the JAs to send the relevant reply message to the MAs.

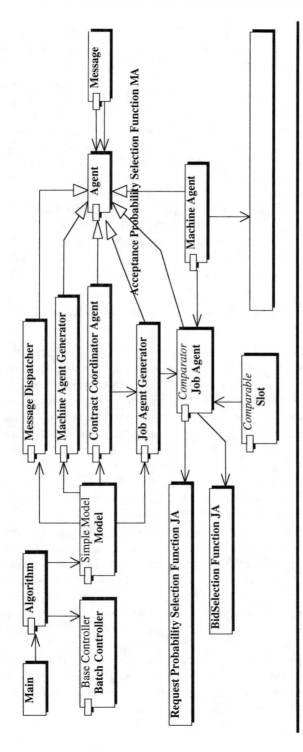

Figure 4.5   Class Diagram of The Parameter Tuning System

■ Activates the offer confirmation or update for the MAs that received an acceptance message at time $t - 1$ from any JA, finally updating the MA private local schedule.

The first bullet point deserves some further comment. In the implementation of the MABS model of the MAS scheduler, the special role of SPG has been assigned to the job generator agent (JGA). This agent is triggered at each time instant by the step( ) method so that it determines any possible new job arrivals according to the relevant random arrival distribution adopted and generates the associated JAs in the MAS. Such a behavior simulates the interaction between the MAS scheduler and the shop-floor planning system. The same JGA returns a termination signal to the simulation engine whenever the maximum number of jobs is reached. In a similar way, a behavior for the machine agent generator (MGA) could be implemented to simulate the interaction of the scheduler with the shop floor in order to take into account possible machine breakdowns.

Executing the MABS-Based Tuning System: Some Results and Conclusions

PS-Bikes' managers opted to analyze the influence of the MAS parameters executing a number of experimental tests with the MABS-based tuning system. Taking into account the typical workload characterizing the production peaks for the framework area, they assumed a maximum number of jobs equal to 300 arriving over time according to an exponential random distribution whose mean interarrival time was set to 16 simulation time instants (where a simulation time instant corresponds to 15 sec in the real system). The maximum number of jobs chosen was also large enough to reproduce the behavior of the shop-floor area under steady stressing conditions. The test was performed assuming that three machines (laser cutters) were initially present in the framework area.

The results generated by the MABS-based tuning system revealed the inadequacy of the number of machines to provide an acceptable performance during peak working conditions; in fact, in such a situation, the reduction of the scheduling cost obtained by the NLOA did not correspond to the performance expected. Thus, PS-Bikes' managers experimented with which changes in the performance might be due to the introduction of additional machines in order to evaluate the trade-off between the cost of these new investments and the expected improvements.

The alternatives evaluated corresponded to adding a fourth and a fifth laser cutter; the tuning process was executed for these alternatives and for the original framework area configuration. Two kinds of results were finally achieved: the "tuned" parameter values for the MAS scheduler to be used in the three configurations (i.e., with three, four, and five machines), and a relative comparison of the performance provided in the same three cases. Although it does not seem particularly interesting to report here the first kind of result, it is worthwhile to show the outcome of the MABS-based tuning system in terms of the relative performance with the three configurations. In particular, Figure 4.6 reports the scatter plots of the three machine configurations and shows how the scheduling performance changes in percentage compared to the initial configuration with only three machines.

For each iteration of the optimization process, PS-Bikes' managers can evaluate the relative differences of the three alternatives compared to the initial three-machine configuration and, in particular, the relative performance when, after a sufficient number of iterations, the NLOA reaches the stop condition. From Figure 4.6, the managers can judge the improvement afforded by the addition of a single laser cutter and the unattractive cost/benefit trade-off in the case of two more machines. A further analysis is possible with the aid of Figure 4.7, which shows the relative behavior of the NLOA. Even in this case, the abscissa reports to the iterations of the optimization process; in the same figure, however, the ordinate denotes the percentage variation yielded with respect to the initial cost for each single configuration.

Contrary to Figure 4.6, no relative comparison of the costs is given in this case, and the three scatters in Figure 4.7 thus start from the same percentage level equal to 1 (100%). The purpose of this latter plot is to highlight the different ability of the NLOA to find a set of values for the scheduler parameters in order to improve the schedule performance. With this insight, PS-Bikes' managers may observe that, with the use of three or four machines (laser cutters), the relative scheduler performance improvement obtained from a parameter-tuning phase may even be inappreciable. As a result, the selection of better parameter values may not be critical in this case. On the other hand, the addition of two machines significantly changes the outcome because, as shown by the plot in the case of five machines, the cost decrease produced by the parameter tuning is quite high.

According to the preceding results, PS-Bikes' managers ultimately make the important choices to introduce a new laser cutter in the framework area swiftly, and to delay the decision about the values to be fixed for the scheduler parameters until after completing further experimental analysis that should take into account other scenarios in addition to peak workload conditions. We can conclude that the use made of the MABS-based tuning system in PS-Bikes went beyond its initial tuning purposes because it proved to be a valuable decision support tool to analyze the costs vs. benefits of different production scenarios.

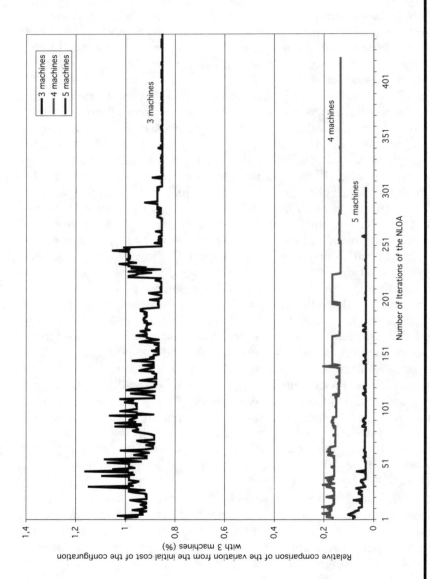

Figure 4.6 Percentage Variations of the Schedule Costs Obtained during the Parameter Tuning Process with Respect to Initial Machine Configuration

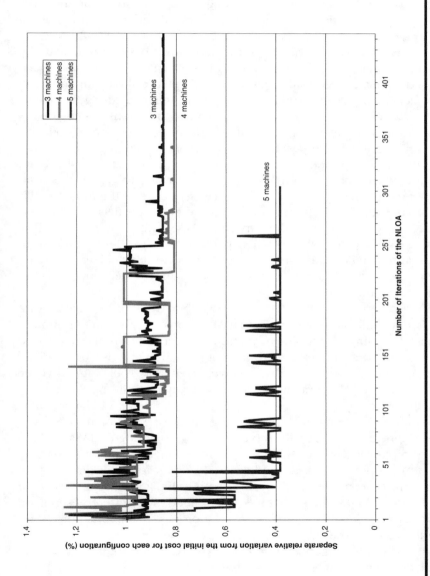

Figure 4.7   Percentage Variations of the Schedule Costs Obtained during the Parameter Tuning Process with Respect to Three Separate Machine Configurations

CONCLUSIONS

Without a doubt, simulation has been one of the most useful computer-based tools for decision-makers in manufacturing; it will continue to play an even more essential role in the era of agile manufacturing. On the other hand, two open questions that have been implicitly addressed throughout this chapter are: how conventional simulation is positioned with respect to the advent of holonic, highly distributed, agent-based manufacturing, and whether new, agent-based simulation paradigms, which have been addressed mainly as MABSs throughout this chapter, will play an important role in facilitating the embedding of agility in a manufacturing company.

Before assessing direct answers to these two questions, some considerations should be made. From what has been reported throughout this chapter, a singular feature of MABSs has emerged: although in conventional simulation a model of the system to be simulated must be built, and agreements with decision-makers must be made to define the complexity of the requested results (and as a consequence of the model itself), in MABSs the gap between the actual MAS and MABS models is not always so wide. Therefore, at times the very same MAS can be used for MABSs. In fact, this is particularly true when parallel and distributed MABSs are used because simulated agents can directly reflect the actual agent behavior, and the parallelization allows the scale-up of the model to the dimensions of the actual systems.

On the other hand, when a parallel distributed multiagent-based simulator is not available when the actual MAS consists of a limited number of agents, the same agent implementations, abstracted from the real world and inserted into a simulated environment (which might also be modeled as an MABS), can work as an MABS as well. Generally speaking, however, when the number of agents in the MAS is elevated or the MAS dynamics is nonetheless complex and takes time to express the MAS behavior fully, the simplification of models, based on MABSs or traditional DES techniques, might be used. For example, business component MASs such as the one proposed by the MAKE-IT approach in the previous chapters could prove more difficult to model the MAS than to run it directly with proper critical scenarios to perform what-if simulations. Conversely, in synthetic social MASs, the same MAS, a simplified modeling representation of it, or a conventional DES model might be proper simulation approaches, depending on the specific case.

Thus, although the general purpose MABS platforms found in literature is simulation environments per se, a novelty that may increasingly be taken up by MABS platforms specific for manufacturing applications is the elimination of the barriers that traditionally separate the worlds of system design, system simulation, and system implementation. As reflected in many experiences reported in the literature, the terms *MAS modeling* and *MAS engineering* are likely to merge all of these tasks, and this unification should be viewed as a strength of novel MAS approaches.

In conclusion, in answer to the two open questions cited earlier, conventional simulation — specifically (parallel and distributed) DES — is still likely to play an important role when a true holonic-based MAS is implemented. However, MABSs should progressively be preferred because they may require less effort in the modeling of simple MASs to the extreme situation in which no relevant differences between the MAS and the related MABS exist. By contrast, in the case of more complex MASs, the definition of a simplified MABS model could also lead to a simpler approach with respect to a conventional DES. Finally, regarding the role of MABSs with respect to manufacturing agility, it should be evident that "if" holonic-based MASs are the appropriate solution to achieve peak performances in an agile manufacturing company, MABSs, too, are a correct solution to simulate behaviors that might help in the study of correct company profiles and configurations. Thus, one of the main of objectives of this book — that is, to demonstrate the cases and the related proper MAS approaches for which the "if" of the previous sentence holds true — should automatically bear out the continued spread and uptake of MABSs.

REFERENCES

1. Bratley, P., Fox, B., and Schrage, L., *A Guide to Simulation*, Springer–Verlag, New York, 1987.
2. Banks, J., Carson, J.S., and Nelson, B.L., *Discrete-Event Simulation*, Prentice Hall, Upper Saddle River, NJ, 1996.
3. Baker, A.D., A survey of factory control algorithms which can be implemented in a multi-agent heterarchy: dispatching, scheduling and pull, *J. Manuf. Syst.*, 37, 297, 1998.
4. Fujimoto, R.M., Parallel discrete event simulation, *Commun. ACM*, 33, 30, 1990.
5. Swain, J.J., Simulation reloaded: sixth biennial survey of discrete-event simulation software tools that empower users to imagine new systems, and study and compare alternative designs, *OR/MS Today*, August, 2003, available at http://www.lionhrtpub.com/orms/orms-8-03/frsurvey.html.
6. Fujimoto, R., *Parallel and Distributed Simulation Systems*, Wiley Interscience, New York, 2000.
7. Lamport, L., Time clocks, and the ordering of events in a distributed system, *Commun. ACM*, 21, 558, 1978.
8. Tropper, C., Parallel discrete-event simulation applications, *J. Parallel Distributed Computing*, 62, 327, 2002.
9. Davidsson, P., Multi agent based simulation: beyond social simulation, in *Multi Agent Based Simulation*, Moss, Scott, Scott J. Moss and Paul Davidsson, eds., Springer–Verlag LNCS series, vol. 1979, 2000.
10. Hare, M. and Deadman, P., Further toward a taxonomy of agent-based simulation models in environmental management, *Math. Computers Simulation*, 64, 25, 2004.

11. Luck, M., McBurney, P., and Preist, C., Agent technology: enabling next generation computing. a roadmap for agent based computing, 2003, available at http://www.agentlink.org/roadmap/index.html.

12. Davidsson, P., Agent based social simulation: a computer science view, *J. Artif. Soc. Social Simulation*, 5, 2002, available at http://jasss.soc.surrey.ac.uk/5/1/7.html.

13. Troitzsch, K.G., Social simulation — origins, prospects, purposes, in *Simulating Social Phenomena*, Conte, R., Hegselmann, R, and Terna, P., Eds, *Lecture Notes in Economics and Mathematical Systems*, Springer, Berlin, Heidelberg, New York, 456, 41, 1997.

14. Moyaux, T., Chaib-draa, B., and D'Amours, S., Agent-based simulation of the amplification of demand variability in a supply chain, in *Proc. Agent Based Simulation 4*, Montpellier, France, 166, 2003.

15. Szirbik, N., Tulba, F., and Wagner, G., Agent-based modelling and simulation of distributed business processes in supply networks, in *Proc. Int. Workshop Modelling Appl. Simulation, MAS2003*, 2–4 Oct. 2003, Bergeggi (I), 100, 2003.

16. Amin, M. and Ballard, D., Defining new markets for intelligent agents, *IEEE IT PRO*, 29, 2000.

17. Parunak, H.V.D., Savit, R., and Riolo, R., Agent-based modeling vs. equation-based modeling: a case study and users' guide, in *Proc. Workshop Multi-Agent Syst. Agent-Based Simulation*, 10, 1998.

18. Wagner, G. and Tulba, F., Agent-oriented modeling and agent-based simulation, in *Proc. 5th Int. Workshop Agent-Oriented Inf. Syst. (AOIS-2003)*, ER2003 Workshops, Springer–Verlag, LNCS, 205, 2003

19. Wagner, G., The agent-object-relationship meta-model: toward a unified view of state and behavior, *Inf. Syst.*, 28, 475, 2003.

20. Gilbert, N. and Bankes, S., Platforms and methods for agent-based modeling, *PNAS*, May 14, 99, 7197, 2002.

21. Swarm, available at http://www.swarm.org/.

22. RePast, an agent based modelling toolkit for Java, available at http://repast.sourceforge.net/.

23. StarLogo, available at http://education.mit.edu/starlogo/.

24. Repenning, A., Ionnidou, A., and Zola, J., AgentSheets: end-user programmable simulations, *J. Artif. Soc. Social Simulation*, Vol. 3, 2000, available at http://www.soc.surrey.ac.uk/JASSS/3/3/forum/1.html.

25. SDML: a strictly declarative modelling language, available at http://sdml.cfpm.org/.

26. The DESIRE Research Programme, available at http://www.cs.vu.nl/vak-groepen/ai/projects/desire/.

27. CORMAS, natural resources and agent-based simulations, available at http://cormas.cirad.fr/indexeng.htm.

28. Vrba, P., MAST: manufacturing agent simulation tool, *EXP — Search in Innovation*, 3, 106, 2003, available at http://exp.telecomitalialab.com.

29. Strader, T.J., Lin, F.R., and Shaw, M.J., Simulation of order fulfillment in divergent assembly supply chains, *J. Artif. Soc. Social Simulation*, 1, 1998, available at http://www.soc.surrey.ac.uk/JASSS/1/2/5.html.

30. Tobias, R. and Hofman, C., Evaluation of free Java-libraries for social-scientific agent based simulation, *J. Artif. Soc. Social Simulation*, 7, 2004, available at http://jasss.soc.surrey.ac.uk/7/1/6.html.

31. Gröbler, A., Stotz, M., and Schieritz, N., A software interface between system dynamics and agent-based simulations – linking Vensim® and RePast®, in *Proc. 21st Int. Conf. Syst. Dynamics Soc.*, New York City, July 20-24 2003, available at http://www.systemdynamics.org/conf2003/proceed/PROCEED.pdf.

32. Bellifemine, F., Caire, G., Poggi, A., and Rimassa, G., JADE. A white paper, *EXP — Search in Innovation*, 3, 6, 2003, available at http://exp.telecomitalialab.com.

33. Bazaraa, M.S. and Shetty, C.M., *Nonlinear Programming: Theory and Algorithms*, John Wiley & Sons, 1979.

34. Hooke, R. and Jeeves, T.A., Direct search solution of numerical and statistical problems, *J. Assoc. Computing Machinery*, 8, 212, 1961.

5

AGENT SYSTEM IMPLEMENTATION

Massimo Cossentino and Luca Sabatucci

This chapter aims to explain how to implement multiagent-based systems. Starting from object-oriented techniques (e.g., UML), great attention is given to the adoption of suitable methodologies for multiagent systems specification (e.g., PASSI), as well as to the importance of standardization (e.g., FIPA) and of the selection of the appropriate languages (e.g., Java) and middleware frameworks to support development and implementation (e.g., JADE). Continuous practical indications referred to the PS-Bikes case study are mentioned throughout the chapter, and more specifically deepened in the second half.

INTRODUCTION

The systematic study of the development of agent systems has a recent history. Little time has elapsed since the scientific world perceived the promise of using the agent paradigm to solve a great variety of problems. This realization prompted many researchers to design, independently, their own infrastructures on which to activate their own agents. The resultant working proposals were often optimal and very efficient for a specific problem domain, but not devoid of some defects. The programming language, communication paradigm, and other technical details generally made these frameworks unsuitable for purposes other than those for which a given approach was originally conceived. The total absence of genuine attention toward the system design and development process (and consequent documentation) often stymied the growth, scalability, and maintenance of these applications. Furthermore, systems were developed without regard to compliance with any standard, thereby creating agents so significantly diverse that they were unable to interact with each other across different frameworks. Now that agent technology has come of age, these solutions, although good for a first experimental phase, are inadequate for the true uptake of this paradigm.

The importance of standardization is such a pivotal issue that an international organization, the Foundation for Intelligent Physical Agents (FIPA), was founded to promote the intelligent agent industry by openly developing specifications supporting interoperability among agents and agent-based applications. A new and very active field, agent-oriented software engineering, is now dealing with the problem of identifying the proper design method for multiagent systems (MASs).

In this chapter we deal with all of these themes — first discussing the key features of FIPA specifications in order to position and define widespread concepts like agent, behavior, and communication in a reference context, and then presenting a complete design process (adopting the PASSI methodology) applied to the PS-Bikes' system case study. The next section examines the standard architecture designed by FIPA for an agent platform and describes the mandatory components that each platform must implement In the third section, using the practical example of the PS-Bikes' system, the fundamentals guiding the implementation of a multiagent system, starting from the initial design down to the code implementation, are illustrated.

The FIPA Abstract Architecture

The work of the FIPA focuses mainly on the definition of the agent platform (AP); this is defined as the physical infrastructure in which agents can be deployed. Most of the standardization work, therefore, concerns the definition of some key points with which an AP must comply. Thanks to these standards, agents living in two or more FIPA-compliant platforms are able to communicate and interoperate with each other. The principal aspects defined by FIPA specifications are:

- The message level, which describes the composition of a message (expressed with the agent communication language), a set of primitive messages with a specific semantic (referring to the speech acts theory [1], and the sequence of speech acts that compose a correct communication (the agent interaction protocol)

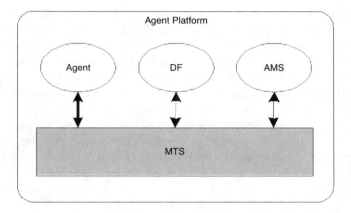

Figure 5.1   Overview of the FIPA Abstract Architecture

- The transport level, which details how a message must be moved from a sender to a receiver
- The service level, which defines the mechanism used by each agent to offer its services and to discover the services offered by other agents in the platform

Architecture Overview

One of the main goals of FIPA specifications is to promote interoperability between agent applications and agent systems; this is achieved by defining the abstract architecture specification. This is a collection of architectural elements that characterize each FIPA-compliant platform. The term "abstract" means that the architecture defines only some functional requirements and is neutral about the technologies used to achieve them.

The agent platform architecture (represented in Figure 5.1) is centered on three mandatory components:

- DF (directory facilitator) component
- AMS (agent management system) component
- MTS (message transport service) component

All of these elements will be examined in detail in the subsections that follow.

Infrastructures for Agent Interactions

The *DF* component of an AP provides the yellow pages service to agents "living" on that platform. It defines the support for agents' collaborations centered on the concept of service defined as an activity that an agent performs on the request of another agent belonging to the same community. Agents may interact with the DF in two different ways: registration and search. To advertise that a specific service is available to the community, the provider agent can register it in the DF with a significant name. Generally, an agent can provide more than one service, each of which is registered in the DF with a different name. An agent has no *a priori* knowledge about the other agents of the system. In order to discover if another can be of any help in reaching its own goal, the agent may search the DF. Consequently, the agent obtains a vector of DF entries; each entry contains the univocal address of an agent of the system that performs that service. Generally speaking, the result is a vector because more than one agent can provide the required service.

The *AMS* is responsible for managing the operation of an AP; the main functionalities of the AMS are the creation, deletion, and life-cycle management of agents. The AMS may support other activities that are not mandatory, e.g., the migration of agents to and from other platforms (mobile agents). The AMS maintains the physical index (AID) of all the agents currently resident on an AP; this index is an address that univocally identifies all the agents of the system.

The *MTS* is generally invisible to agents and their developers. It provides a mechanism for delivering messages among agents within a platform and to agents resident on other platforms. Messages are coded in a standard structure composed of an envelope and a payload. The envelope contains transport information needed for the correct delivery of the message. Transport information could specify a network protocol like HTTP or SMTP and the address of the agent if it is reachable using that protocol (something like *www.mysite.net/abc* or *agent-name@host.domain.org*). The payload record is coded in a language called agent communication language (ACL), and it contains the information content that is to be delivered.

Agent Social Relationships

Social relationships are among the most important characteristics of agents. An MAS is composed of a number of autonomous and interacting agents and is frequently represented as a well-organized society of individuals. In this context, each agent has its personal goals and plays one or more different roles during its life to interact with other community members.

Agents interact through messages only and, most commonly, their interaction is made up of a series of messages, thus composing what is defined as a conversation. It is correct to think about an agent interaction as a conversation rather then one simple message. A conversation, and specifically an FIPA conversation, is essentially composed of one or more messages. As already mentioned, each message needs a transport infrastructure in order to be delivered. This allows the effective implementation of a conversation but does not ensure any usefulness for it. In order to add a semantic value, five important concepts must be adhered to (see Figure 5.2): ontology; content, content language, communicative act, and agent interaction protocol (AIP).

Modeling the Communication Semantic with an Ontology

Ontology is a representation of the categories that exist in a specific domain; it is a vocabulary used to describe the terms and the relationships among them with a subject matter. Ontology allows the specification of the types of terms an agent may handle and what type of manipulation and reasoning it is able to perform on them. Referring to the same ontology, two agents can interact without the risk of a misunderstanding. They refer to the same set of concepts and, if they adopt the same (content) language, the communication will be meaningful for both of them. Conversely, the lack of a common ontology introduces the risk that a term used by an agent with some specific significance will be interpreted by another in a different way, thereby jeopardizing agents' interaction and the entire system's performance.

Ontology defines the meaning of categories and the relationship among them, but, in order to manage it, agents need a language that can represent ontology structure and content. In many approaches, the ontology structure is composed of three kinds of elements (concepts, predicates, and actions) and the associations among them. Many authors have dealt with the representation of ontology using unified modeling language (UML) [2, 3]. In this chapter, we will adopt the PASSI [4, 5] notation that uses a UML class diagram. Concepts, predicates, and actions are represented as classes characterized by a specific stereotype. Figure 5.3 reports a PASSI diagram representing a portion of the ontology designed for a PS-Bikes MAS.

As an example, the *Order* class (Figure 5.3) represents a concept of the ontology; a concept stands for one of the categories of the specific domain and, in this example, *Order* represents the order issued by a customer for receiving some bicycles. It has some attributes, e.g., the *delivery_date*, which is the delivery date requested by the customer for the ordered goods. A concept may be related to other concepts; for example, an order is composed of one or more *OrderStock* (i.e., the number of bicycles of a certain model specified in the order). A concept may extend another concept, inheriting all the attributes and relationships of its superconcept. For example, a *Customer* is a specific *Company* with

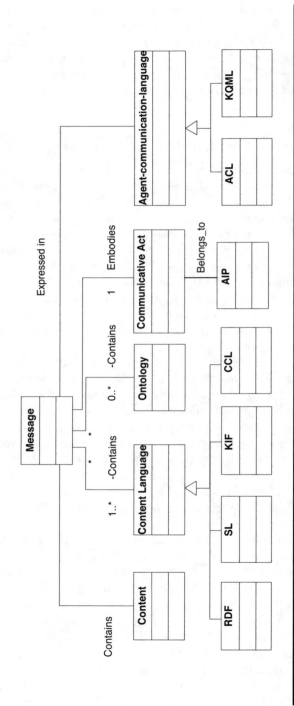

Figure 5.2 Structural Diagram Illustrating the Elements Constituting an FIPA Message and Relationships among Them

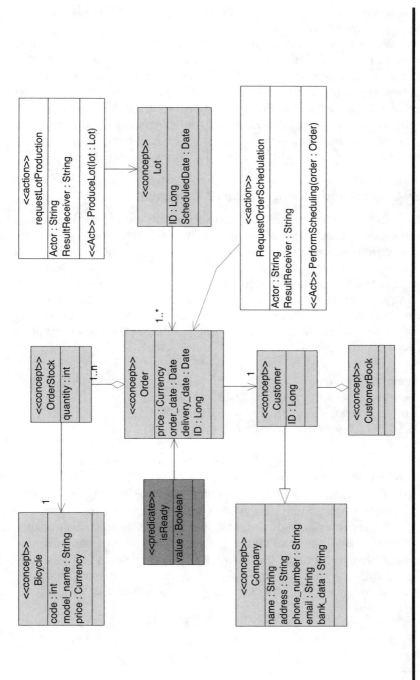

Figure 5.3 Example of Ontology Expressed Using a UML Class Diagram (Domain Ontology Description Diagram of the PASSI Methodology)

some supplementary characteristics (the *ID* attribute used to identify it in the bicycle production company).

A predicate represents a particular statement or belief surrounding some concept, as in the case of the *isReady* predicate shown in Figure 5.3. This is used to announce that some specific *Order* is ready to be delivered.

An action indicates the type of operation that can be performed on elements of the ontology, thus possibly provoking some changes to the internal knowledge of the agent. *RequestOrderSchedulation*, in Figure 5.3, is an example of an action specifying the request from one agent to another to schedule the production of bikes for some specific order.

Message Content and Message Content Language

The MTS is the architectural level of a platform that performs the routing of a message from the sender to the receiver whether they are in the same or in different platforms. The life cycle of a message from its initial creation by the sender to its reading by the receiver agent is illustrated in Figure 5.4. The basic information delivered by a message is taken from the ontology of the sender agent; it could be a concept, a predicate, or an action. The message content (that refers an element of the ontology) is expressed by the agent using a content language. FIPA specifications include four languages:

- Semantic language (SL)
- Constraint choice language (CCL)
- Knowledge interchange format (KIF)
- Resource description framework (RDF)

These are born in different contexts and represent the solutions adopted in specific approaches or by some communities; each of them has its specific domain in which it is preferable. The RDF language was created for Web applications, but, as previously alluded, it has proved to be optimal for representing an ontology for many different applications. It is frequently used, alternatively to SL, as the content language of messages exchanged among FIPA agents. The other two languages, CCL and KIF, were developed for artificial intelligence applications; they are very powerful at expressing actions and predicates, but they come with a complex grammar.

The RDF language enjoys very widespread use because (1) it is a W3C* and an FIPA† standard; (2) it has quite a simple syntax; and (3) it allows

---

* World Wide Web Consortium RDF specifications: http://www.w3.org/RDF/
† FIPA RDF specifications: http://www.fipa.org/specs/fipa00011/

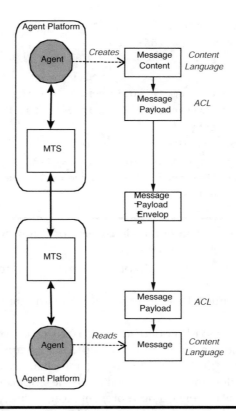

Figure 5.4 Transformations of a Message during Its Life

a number of possible representations (e.g., it also exists in the form of an XML specification). The RDF description (expressed using XML) of the ontology element *Bicycle* illustrated in Figure 5.3 is shown in Figure 5.5.

Once the message content is expressed in a content language, it is necessary to encapsulate it into a structure called message payload. This structure is coded in a specific ACL that includes several other message parameters, the most relevant of which are:

■ Performative: type of communicative acts (inform, request, agree ...), which depends on the AIP
■ Sender: ID of the agent playing the sender role in the communication
■ Receiver: ID of the agent playing the participant role in the communication
■ Content: the already discussed message content (express in a content language)
■ Language: language used for the message content

```
<rdfs:Class rdf:ID="Bicycle">
        <rdf:type rdf:resource="rdfsx:concept"/>
</rdfs:Class>
<rdf:Property ID="Bicycle.model_name">
        <rdfs:domain rdf:resource="#Bicycle"/>
        <rdfs:range rdf:resource="rdfsx:String"/>
</rdf:Property>
<rdf:Property ID="Bicycle.price">
        <rdfs:domain rdf:resource="#Bicycle"/>
        <rdfs:range rdf:resource="#Currency"/>
</rdf:Property>
```

Figure 5.5   RDF Description of Bicycle Element of Ontology Shown in Figure 3

■ Ontology: name of the ontology element reported in the message content
■ Protocol: name of the aip used in the communication

The message payload, coded in ACL, is received by the MTS of the platform where the sender agent is located. MTS encapsulates the payload into an envelope including the transport information needed to deliver the message: sender and receiver transport descriptions, plus additional information such as the encoding representation, security-related data, and whatever else needs to be visible to the MTS. The transport descriptions describe the transport protocol to be used (IIOP, HTTP, and WAP are examples of such protocols) and the physical address (e.g., an IP address) to which the message must be delivered.

Agent Interaction Protocols

The FIPA abstract architecture places a great deal of importance on the interaction rules of agent conversations. These have been formalized primarily through two concepts: the communicative act and the AIP (also known simply as "protocol" in this context). According to the FIPA directive, each conversation must respect a protocol and must be made up of communicative acts (see also Figure 5.2). A *communicative act* is a way to associate a predefined semantic to the content of a message so that it can be univocally understood by agents. The FIPA is responsible for maintaining a consistent list of communicative acts. Some examples of communicative acts are illustrated in Figure 5.6; they are *request*; *refuse*; *agree*; *inform*; and *failure*.

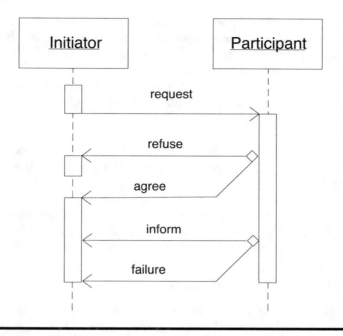

Figure 5.6 FIPA Request Interacting Protocol

A *protocol* univocally defines which communicative acts may be used in a conversation and the order in which the related messages must be sent to give the proper meaning to the communication. Therefore, a protocol compels the use of determined messages with a specific semantic according to a specific sequence. When an agent starts a conversation with another agent it must specify a protocol; a conversation without a protocol is not possible. If a message does not respect the rules of the protocol or violates the prescribed order, then the conversation fails.

Until now, FIPA specifications have used AUML diagrams [6,7] to describe protocols. This diagram is a modified version of the UML sequence diagram. The FIPA request interaction protocol is illustrated in Figure 5.6. This may be used when one agent (the initiator) asks another (the participant) to perform some kind of action.

To start the conversation, the initiator sends a *request* communication act. The content of the message is a description of the action to be performed, constructed in a language the receiver understands; if there is a common ontology, the content may be an ontology action (as described in the previous paragraph).

The participant processes the incoming *request* and decides whether to accept or refuse it. The receiving agent makes a decision on the basis of a type of reasoning as could be expected given the principle of autonomy of agents. If the participant agent agrees to perform the requested action, then it replies with an *agree* message; otherwise a *refuse* message is sent (the possibility of sending an *agree* or *refuse* response is represented in Figure 5.6 by the diamond).

Once the request has been accepted, the participant must fulfill the action and, according to the result obtained, reply with one of the following communicative acts:

■ A *failure* message to notify that the action was not completed for some reason; this motivation is usually reported in the content of the message

■ An *inform* message to communicate that it successfully carried on the action to be done; some information on the action results may be reported in the content of the message (e.g., a link to a Web site selected according to criteria passed on by the initiator agent)

JADE: an Implementation Platform

The FIPA describes an abstract architecture that cannot be directly implemented; because the main focus of these specifications regards agent interoperability, not many details are provided on the platform implementation aspects. On this basis, a great number of different solutions have been proposed over the last years, a list of which can be found on the FIPA Web site. Among the most widely used are FIPA-OS, JADE, and Zeus. In this subsection, the JADE AP is briefly analyzed in order to illustrate some of its specific implementation details.

JADE (Java agent development framework) [8] was completely developed in Java language by Telecom Italia Lab with the collaboration of the University of Parma. The JADE platform has many interesting features; one of these is the support that it provides for agent mobility, which allows its use for the creation of distributed applications in which mobility plays an important role (e.g., searching).

A JADE agent is based on a class that extends the *Agent* superclass (a UML class diagram representing the *Administration* agent from the bicycle case study reported in the next subsection is shown in Figure 5.7). The agent class usually contains a constructor (required by Java and, by convention, in JADE used to initialize data structures) and the *setup* method, which, automatically invoked by the platform once the constructor ends, is often used to begin the agent activity. An agent can be instantiated only by the platform; when this happens, a univocal ID is assigned to the agent and the constructor, followed by the *setup* method, is executed. Often, the developer uses the constructor to initialize the agent's data structures and the *setup* method to start the activity of its agent.

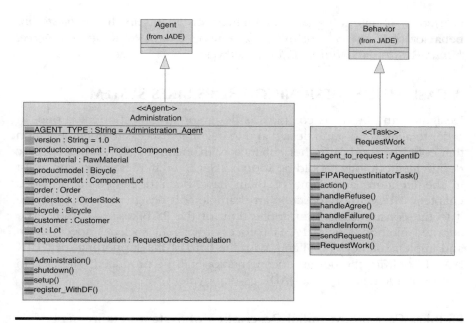

Figure 5.7   Structure of a Jade Agent with a Behavior

Another method automatically invoked by the platform is *shutdown*, which arises when an agent is about to terminate. It contains the code needed to conclude the agent's activities properly and to reallocate the assigned resources. The JADE *Agent* class (the mother class of all the agents) already provides such a method and, in most cases, this is sufficient to shut down the agent successfully.

Agent activities are typically not described in its base class methods, but are located in some subclasses called *behaviors*. A behavior represents the atomic element of decomposition of the agent's tasks. Operations needed to reach a goal of the agent are partitioned among its behaviors. For instance, communication with another agent is delegated to a specific behavior (an example is the *RequestWork* class shown in Figure 5.7). Concretely, a behavior is a class that extends a JADE superclass called *Behavior*. As seen for the agent base class, a template structure exits for behavior classes. All the behaviors must contain an *action* method. Like the *setup* method, *action* is automatically invoked by the platform, after which the class constructor method is completed; the use is the same but at the behavior level (i.e., it is used to start the operations related to that behavior).

Obviously, a behavior class can contain several methods; a communication behavior is usually made up of a set of methods in order to catch all the incoming messages of a specific protocol. For instance, if a behavior is used to initiate a *Request* communication [9] (as in the *RequestWork* behavior of Figure 5.7), it must contain the *handleRefuse*, *handleAgree*, *handleFailure*, and *handleInform* methods.

A CASE STUDY: DESIGNING THE PS-BIKES SYSTEM

Designing an MAS is as complex as designing an object-oriented one. In order to achieve a sound design and to guarantee access to documentation that could be used to further enhance or maintain the software, a specific design methodology should be adopted. Several different approaches exist in the literature and some of them have been discussed in previous chapters. We will now describe an example of a design process, applying it to the construction of an application for the PS-Bikes case study. The adopted methodology is PASSI (process for agent societies specification and implementation) [4, 5] and, with the help of the supporting tool, PTK (PASSI ToolKit), the design documentation will be produced. The system will be implemented using JADE as deployment AP.

PS-Bikes Case Study: Initial Description of System Requirements

The first phase of the design in most methodologies entails the elicitation and analysis of requirements. A requirement is a feature that the system must exhibit; it can be functional, such as service, or nonfunctional, such as a constraint or a performance issue. In UML [10] (functional) requirements are described with use case diagrams. According to UML [11], a use case represents a coherent unit of functionality provided by a system, subsystem, or class, as manifested by sequences of messages exchanged throughout the system (subsystem, class) and one or more outside interactors (called actors), together with actions performed by the system (subsystem, class). An actor defines a coherent set of roles that users of an entity can play when interacting with the entity.

In Figure 5.8, a use case diagram depicts the functionalities of a portion of the PS-Bikes' system and the interactions with two actors: the customer department and the production supervisor.

The company organizes its production on the basis of the received orders. The customers are wholesalers and retailers of sporting goods; they interact with a figure called the customer department represented by an actor (a stick figure) in the diagram. When a customer wants to place an order for some bicycles, he contacts the customer department directly (e.g., sending the order by fax); using a graphical interface the customer department employee that receives the customer order may introduce the data into the system. This functionality is represented by the "order acquisition" use case. The "customer data management" functionality

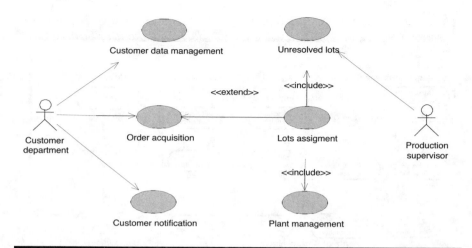

Figure 5.8   A Portion of Use Case Diagram Representing Functionalities of PS-Bikes System

allows the company to maintain an archive of customers. The administration department generates plans for the production phases of the two plants on the basis of forecasts of the demands and customers' orders. When an order is placed by the customer department, it must be composed in lots and its production assigned to a specific plant. These operations are represented by the "lot assignment" and "plant management" use cases. The person responsible for interacting with the lot scheduler is the *production supervisor.*

DESIGNING THE SOLUTION WITH PASSI

It is well known that code production is a complex activity, and the agent-oriented paradigm does not ignore this hurdle. A methodology to design and implement MASs is a prerequisite approach to simplify this task. The PASSI methodology is a step-by-step, requirements-to-code methodology for designing and developing multiagent societies. It integrates design models and concepts from OO software engineering and artificial intelligence approaches using the UML notation with some extensions.

As already mentioned, the methodology is supported by PTK (PASSI Toolkit), a Rational Rose plug-in, and also by a repository of patterns for agents. These tools are very useful in the design and development of the MAS because they introduce a level of automation into the process, thus enhancing the designer's productivity. This is particularly effective when entire portions of the model are taken from the patterns repository; this reuse, performed during the design phase, also affects the coding activity

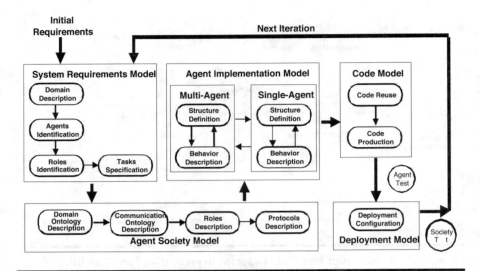

Figure 5.9   Different Steps and Models of the PASSI Design Process

because a significant portion of code is automatically generated starting from the pattern structure.

In the following subsections, the PASSI methodology is synthetically analyzed in order to illustrate how a methodology specifically conceived for multiagent systems can support and simplify the designer's work. The methodology is applied to the design of a system for the PS-Bikes case study.

The PASSI Methodology

PASSI is composed of five models (Figure 5.9) regarding the different abstraction levels of the process:

- ■ System requirements model. The initial part of this model is similar to other common object-oriented methodologies (requirements analysis). An agent-based solution to the problem is thus drafted.
- ■ Agent society model. This describes the details of the system solution in terms of agent society concepts like ontology, communications, and roles.
- ■ Agent implementation model. The previous models are used to obtain a detailed description of the agent society in terms of structure and behavior that can be used to produce the code of the system.
- ■ Code model. In order to streamline and speed up the development of a new system, code is partially obtained from the application of patterns. A conventional code completion activity is then carried out.

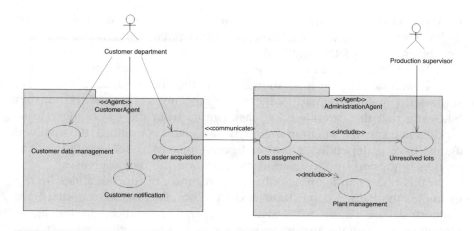

Figure 5.10 Portion of Agent Identification Diagram for PS-Bikes Case Study

- Deployment model. Mobile agents require that specific attention be paid to the specification of their needs in terms of software environments (e.g., libraries available in the host platform) and hardware capabilities and performance (e.g., amount of available network bandwidth); these are the issues defined in the deployment model.

The System Requirements Model

The system requirements model is a model of the system requirements in terms of agency and purpose. The methodology is driven by use case and starts with the requirements analysis, in which the designer models the system as a set of use case diagrams. Some of these diagrams, the domain (requirements) description diagrams, are drawn to represent the actors and the use cases identified for the system. Figure 5.8 reports some of the use cases of our PS-Bikes' system. In this kind of diagram, the designer can identify the agents that will populate the solution. In PASSI, each agent receives the responsibility for a part of the functionalities of the whole system; this is represented in a use case diagram (called agent identification diagram) by grouping some of the use cases within a package and giving it the name of the agent.

Figure 5.10 depicts a portion of the Agent Identification diagram for the PS-Bikes' system. It describes only two agents, the *Customer* and the *Administration*; these are displayed as two packages containing some use cases from Figure 5.8. Each agent is responsible for accomplishing the functionalities associated with the use cases included in its package. For example the *Customer* agent responsibilities include: *Customer Data Man-*

agement, *Order Acquisition*, and *Customer Notification*. All of these have a direct interaction with the *Customer department* actor that represents one of the users of the application.

When two use cases are assigned to different agents and are related by an *include* relationship (showing that the included use case offers some kind of functionality to the including one) or *extend* relationship (showing that the extended use case profits from the extending one to tackle some specific situation triggered by a guard condition), the involved agents have a dependency and will communicate to achieve the collaboration requested by the relationship between the two use cases.

In this phase, an agent is only an aggregation of functionalities. In the example, the *Order Acquisition* and the *Lots Assignment* use cases are connected (see Figure 5.10) with an *extend* association. In the agent identification diagram, this turns into a *communicate* relationship (representing an agent conversation) between the two agents.

When all the agents are identified, the next step is to explore the scenarios in which they are involved. This is done using a set of UML sequence diagrams; in these diagrams, each agent may be involved in many activities and may appear more than once in each single scenario; this means that an agent plays more than one role in that scenario. The identification of agent roles is one of the main outcomes of these diagrams, which are therefore called role identification diagrams in PASSI. An example of a role identification diagram is shown in Figure 5.11. Here, the *Customer* agent appears twice: in the first instance, it searches for information about a customer in the company database (role *CustomerDB*) and then, in the second, it archives a new customer's order (role *Order-Management*).

The last step of this first model (the system requirements model) is to begin to describe the dynamic behavior of each agent. This phase is performed with a set of task specification diagrams (one for each identified agent). According to FIPA definitions [12], a task is "the observable effect of an operation or an event, including its results. It specifies the computation that generates the effects of the behavioral feature." Starting from this definition, PASSI considers a task as an entity that is somehow similar to the *Behavior* defined in the JADE agent structure. The task specification diagram is a UML activity diagram representing agents in a swim lane and their tasks as activities. Each diagram is drawn to detail one agent and only two swim lanes are present in it (see Figure 5.12): the right-hand one contains a collection of activities symbolizing the current agent's tasks, while the left-hand one reports some activities from other agents involved in interactions with this specific agent.

An example of a task specification diagram for the *Administration* agent is illustrated in Figure 5.12. This agent is involved in the introduction

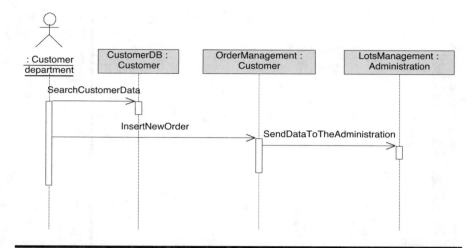

Figure 5.11   Role Identification Diagram for "Insert New Order" Scenario

of a new order from a *Customer* agent. It receives this communication with the *OrderListener* task. After that, the agent plans the bike production with the *PlannerTask* and *RequestWork* tasks. The *SupervisorGUI* task is activated if a problem is found in the planning phase; the task is responsible for notifying the production of the need to adjust the plan manually.

The Agent Society Model

The next PASSI model is the agent society model that represents social interactions and dependencies among agents involved in the solution. This model is composed of four phases:

- Domain ontology description: the domain is explored and its distinguishing concepts are identified together with actions and propositions related to them
- Communication ontology description: used to detail agent communications in terms of ontology, content language, and interaction protocol
- Roles description: consisting of a diagram representing agents with their roles, the tasks involved in those roles, and the dependencies among agents and roles in terms of resources to be shared and services to be provided
- Protocols description: constituting a phase that is frequently skipped by the designer; it is necessary to define a new protocol only if the existing FIPA protocols are insufficient to model the specific communication (which happens rarely)

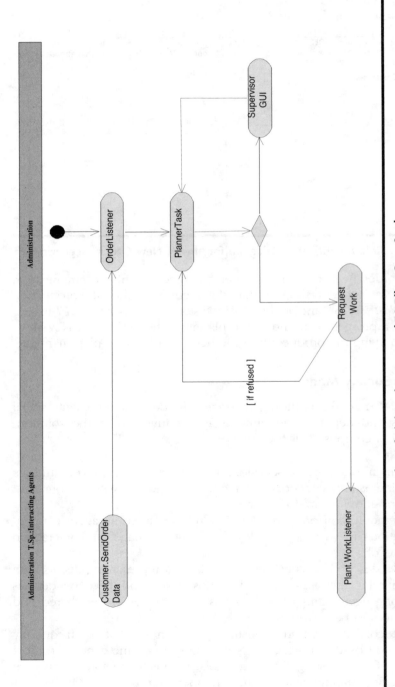

Figure 5.12 Task Specification Diagram for Administration Agent of PS-Bikes Case Study

In the PASSI methodology the design of ontology is performed in the domain ontology description (DOD) phase and a class diagram is used. Several works can be found in the literature about the use of UML for modeling ontology [6, 7, 13]. Figure 5.3 reports an example of a PASSI DOD diagram; it describes the ontology in terms of concepts (categories, entities of the domain); predicates (assertions on properties of concepts); and actions (performed in the domain). This diagram represents an XML schema that is useful to obtain a resource description framework (RDF) encoding of the ontological structure. We have adopted RDF to represent our ontologies because it is part of the W3C [14] as well as FIPA (FIPA RDF content language) [15] specifications.

In Figure 5.3, the PS-Bikes system ontology is described by classes and their relationships. Elements of the ontology are related using three UML standard relationships:

- Generalization permits the "generalize" relation between two entities — one of the essential operators for constructing an ontology.
- Association models the existence of some kind of logical relationship between two entities and allows specifying the role of the involved entities in order to clarify the structure.
- Aggregation can be used to construct sets in which value restrictions can be explicitly specified; in the W3C RDF specification, three types of container objects are enumerated: the bag (an unordered list of resources); the sequence (an ordered list of resources); and the alternative (a list of alternative values of a property). For our purposes, we consider a bag as an aggregation without an explicit restriction and a sequence as qualified by the *ordered* attribute; the alternative is identified with the *only_one* attribute of the relationship.

The example (Figure 5.3) shows that each *Order* concept is characterized by a *price, order_date, delivery_date*, and *ID*. Each order aggregates several *OrderStocks*, each of which describes the number of bikes of a specific type that are part of the order. The bicycle model is described in the homonymous concept. One agent can ask another if an order has been completed, and this instance is stated by the Boolean value of the *isReady* predicate. The *ScheduleManifacturing* action introduces the order (and therefore the specified number of bicycles) in the manufacturing scheduling of the different machine tools.

The communication ontology description (COD) (Figure 5.13) is a representation of the agents' (social) interactions; this is a class diagram that shows all agents and all their interactions (lines connecting agents). In designing this diagram, we start from the results of the agent identifi-

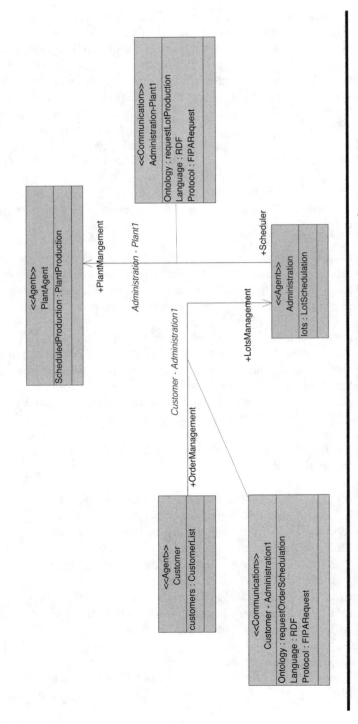

Figure 5.13 Communication Ontology Description (COD) Diagram for PS-Bikes Case Study

cation (AID) phase. A class is introduced for each identified agent, and an association is then introduced for each communication between two agents (ignoring for the moment distinctions about agents' roles). Clearly, it is also important to introduce the proper data structure (coming from the entities described in the DOD) in each agent in order to store the exchanged data.

The association line that represents each communication is drawn from the initiator of the conversation to the other agent (participant) as can be deduced from the description of their interaction performed in the role identification (RID) phase. As already mentioned, each communication is characterized by three attributes, which we group into an association class. This is the characterization of the communication itself (a communication with different ontology, language, or protocol is certainly different from this one); its knowledge is used to refer this communication uniquely (which can have, obviously, several instances at runtime because it may arise more than once). Roles played by agents in the interaction (as derived from the RID diagrams) are reported at the beginning and the end of the association line.

Figure 5.13 illustrates the communication between the *Customer* and *Administration* agents (the unique communication name is: *Customer-Administration1*). The first initiates the interaction in order to ask the other about the production scheduling of an order for some bikes. The referred ontology is an action (*requestOrderSchedulation*) and the inter-action protocol is the FIPA request that is dedicated to dealing with requests for some kind of service. RDF is the content language.

The FIPA methodology glossary [12] defines a role as "a portion of the social behavior of an agent that is characterized by some specificity such as a goal, a set of attributes (for example responsibilities, permissions, activities, and protocols), or providing a functionality/service." In PASSI, roles are initially identified in the previously discussed AID diagrams. Their definition is completed in the role description (RD) diagram, i.e., a UML class diagram in which classes are used to represent roles. Agents are represented by packages containing classes of roles (see Figure 5.14). Each role is achieved by grouping several elementary tasks into a resulting complex behavior; for this reason, tasks are shown in the operation compartment of each role's class. During its life, an agent can take on several different roles, and this dynamic evolution in its behavior is represented by a dashed line with the name [ROLE CHANGE] that connects its different roles in the expected order. Conversations between roles are indicated by solid lines (as depicted in the COD), using exactly the same relationships names.

We have also considered dependencies between agents. Because agents are autonomous and may refuse to provide a service or a resource

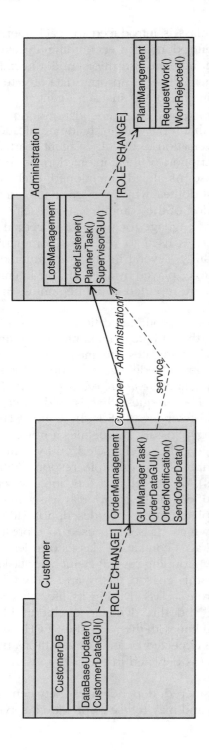

Figure 5.14 Role Description (RD) Diagram for PS-Bikes Case Study

to another, the design needs a schema that expresses such matters and explores alternative ways to achieve goals. In order to realize such a schema, some additional relationships that express the following kinds of dependency have been introduced into the roles description diagram:

- Service dependency: one role depends on another to bring about a goal (indicated by a dashed line with the *service* stereotype)
- Resource dependency: one role depends on another for the availability of an entity (indicated by a dashed line with the *resource* stereotype)
- Soft-service and soft-resource dependency: the requested service or resource is helpful or desirable, but not essential to bring about a role's goal (indicated by a dashed line with the *soft-service* and *soft-resource* stereotypes)

In the example of Figure 5.14, the *Customer* agent plays the *CustomerDB* role while dealing with the customer data and the *OrderManagement* role while managing customer orders. We can see that several tasks are involved in the exploitation of the second role (e.g., graphical interfaces like *OrderDataGUI* are used to interact with the user that introduces the customer order data). We can also note that this agent initially plays a role related to the compilation of the customer data archive and then changes its vocation (*Role Change* relationship) toward order-oriented operations. The communication with the *Administration* agent already discussed in the COD diagram (Figure 5.13) is also reported in order to simplify the analysis of the interactions among the different roles.

As seen in the DOD phase and as specified by the FIPA architecture, a protocol is used for each communication. All of them are standard FIPA protocols in this case study. Usually, the related documentation is given in the form of AUML sequence diagrams [16]; thus, designers do not need to specify protocols on their own. In some cases, however, existing FIPA protocols are not adequate. If this happens, some specific protocols must be properly designed (protocol description phase); this can be done using the same FIPA documentation's approach (with an AUML sequence diagram as in Figure 5.6).

The Agent Implementation Model

The agent implementation model is a model of the solution architecture composed of two different phases, each performed at the multi- and single-agent level of abstraction. The multiagent level deals with the agent society and is therefore detailed to a low degree as regards the agent implementation specifications; however, it fittingly documents the overall structure of the system (behaviors of each agent, communications, etc.). The single-agent level of abstraction focuses on the implementation details of each agent and specifies whatever is needed in order to prepare the coding phase. The two phases are:

- Agent structure definition (ASD) uses conventional class diagrams to describe the structure of solution agent classes.
- Agent behavior description (ABD) uses activity diagrams or state-charts to describe the behavior of individual agents.

This model is characterized by an iterative process and, specifically, by a double level of iteration (see the agent implementation model box in Figure 5.9). This model needs to be viewed as being composed of two views: the multiagent and single-agent views related by two iterations. The outer level of iteration concerns the dependencies between these two views. In each we can find an ASD (representing the agents' structures at the social or inner-agent granularity) and an ABD (describing the agents' behaviors again from the social or single-agent perspective). An inner level of iteration takes place at the multiagent and single-agent views and concerns the dependencies between the structural and behavioral matters. As a consequence of this double level of iteration, the agent implementation model is composed of two steps (ASD and ABD), but still yields four kinds of diagrams when the multi- and single-agent views are taken into account.

In the multiagent structure definition (MASD) diagram, attention is centered on the general architecture of the system. The MASD is an overview of the results obtained from the previous phases from the structural point of view. In this diagram (Figure 5.15), agents are represented as classes with their behaviors in the operations compartments; attributes specify the agent knowledge. Building this diagram is not an effort for the designer because PTK (the tool that supports the design with the PASSI methodology) automatically builds it using information coming from previous diagrams.

At this point, a new diagram, the single-agent structure definition (SASD) diagram, is drawn for each agent in order to explore its internal composition and all of its tasks at a level of detail sufficient to generate the implementation code. This diagram is a UML class diagram and reports the agent main class and each agent task as a class, resembling the structure of the most common AP specifications (Jade [8], FIPA-OS [17]). At this point, we set up attributes and methods of the agent class (e.g., the constructor and the shutdown method required by the FIPA-OS platform or just the constructor in JADE) and the task classes (e.g., the methods

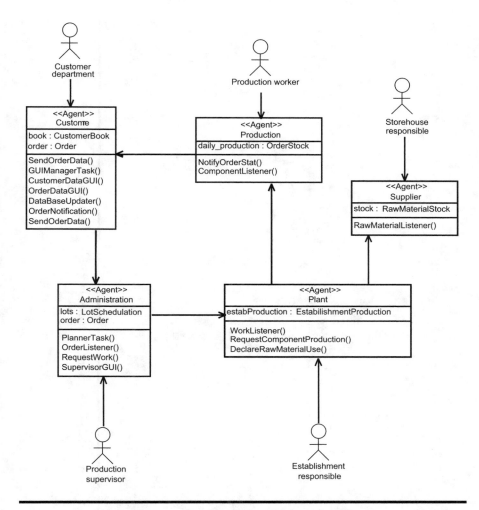

Figure 5.15 Multiagent Structure Definition Diagram for PS-Bikes Case Study

required to deal with communication events when the agent receives or sends a communicative act).

An example of an SASD diagram is illustrated in Figure 5.16 and describes the internal structure of the *Customer* agent of the PS-Bikes case study to be implemented in the JADE platform. The *Customer* main class is derived from the *Agent* base class of JADE. Among its attributes is *AGENT_TYPE,* which usually contains the name of the agent type (*Customer* in this case) and, in the operations compartment, the *register_WithDF* method that contains the code necessary to register with the yellow pages service of the platform (*Directory Facilitator*).

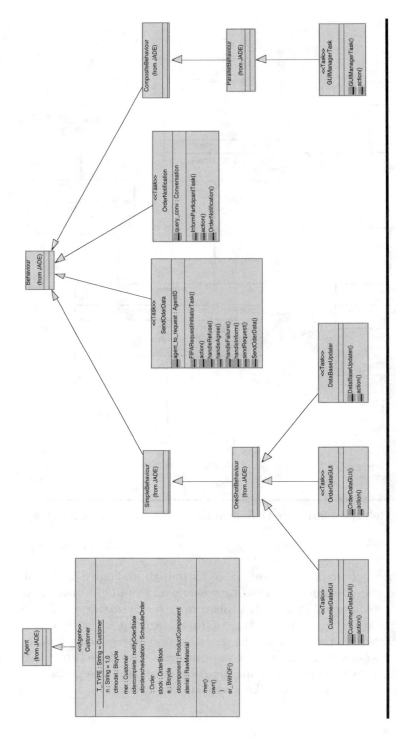

Figure 5.16 Single-Agent Structure Definition (SASD) Diagram for *Customer* in PS-Bikes Case Study.

As regards the agent's tasks (called *Behaviors* in JADE), we can consider *SendOrderData* and *OrderNotification,* which are represented as two classes extending the JADE *Behavior* super class, and whose duties entail dealing with the agent communications (as can be seen in Figure 5.15, this agent has relationships with the *Production* and *Administration* agents). For example, *SendOrderData* adopts a "request" protocol to delegate the *Administration* to take care of the introduction of a new order in the manufacturing schedule.

A different structure is proposed for *CustomerDataGUI, OrderDataGUI* and *DataBaseUpdater,* which are inherited from the JADE *OneShotBehavior* (a behavior that performs a single operation and then terminates its existence). This kind of solution is a valid option for controlling graphical interfaces, i.e., once the interaction with the user is completed, the behavior does not need to remain active.

The agent behavior at the multiagent level is described by the multiagent behavior description (MABD) diagram. This is a UML activity diagram used to illustrate the dynamics of the system during the agent's life. Figure 5.17 reports an example of MABD; it illustrates the activities occurring during the *Request* communication between the *Customer* and *Administration* agents. In the diagram, all the involved classes (of agents and tasks) are represented with swim lanes (such as *Customer* and *Customer.SendOrderData*), while operations are displayed as an activity (rectangles with rounded corners, like *SendOrderData.PrepareRequest,* which is the constructor method of the *SendOrderData* behavior in Figure 5.17). In these diagrams, transitions among activities indicate an event as a method invocation (if relating activities in the same swim lane); a new behavior instantiation (if relating activities of the same agent but in different swim lanes); or a message (if two different agents are involved). The communication described in the example initiates a request message and then, according to a decision process (not described), the *Administration* agent replies with a *refuse* or *agree* message. Each message is detailed with the communication name and the communicative act.

The single-agent behavior description (SASD) is the last phase of the agent implementation model. The approach used in this activity is quite common. The aim of this phase is to produce a design of the inner part of methods introduced in the SASD diagrams in order to prepare their implementation. The designer is free to describe these features as he or she sees most fitting and appropriate (e.g., using flow charts, state diagrams, or semiformal text descriptions). It should be noted that, because in many instances operations performed according to a method are not complex enough to justify so much attention, a textual description is often sufficient.

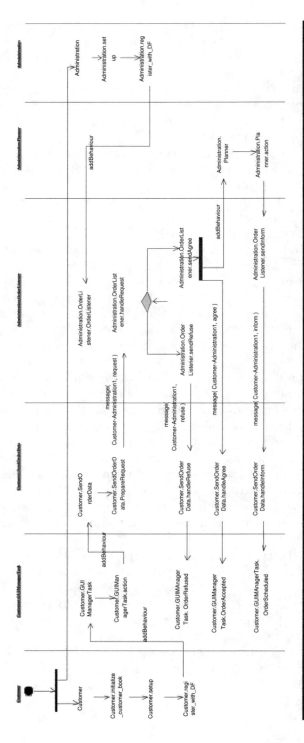

Figure 5.17 Multiagent Behavior Description (M.A.B.D.) Diagram Used to Describe Interaction of Two Agents during FIPA Request Communication

The Code Model

The code model is a model of the solution at the code level. In this phase, the developer is aided by a tool (AgentFactory) developed in the order to grant the code reuse. AgentFactory may work inside PTK or as a standalone application. Its key feature is that it allows the easy construction of a substantial part of an MAS reusing elements of its pattern repository (specifically realized to solve agent-oriented problems and therefore different from a common object-oriented one).

An agent pattern, according to the PASSI conception, derives from object-oriented design patterns [18] and describes a tested solution for a recurrent design problem. This pattern [19, 20] is presented as a set of diagrams of the PASSI methodology, each describing the different aspects of the problem at different abstraction levels and covering one or more phases of the design process. Typically, diagrams used to describe a pattern are classified in one of two categories: structural or behavioral. The most common diagrams used in the pattern description are the task specification, DOD, COD, SASD, and MABD. Starting from these representations and from a description of the solution with an XML-based metalanguage, AgentFactory can instantiate the implementation code for the FIPA-OS and JADE platforms. Obviously, the code generation engine also considers the needs emerging from the composition of different parts to create a complex agent structure and can solve all the ensuing problems.

Communication patterns are among the most frequently used by the AgentFactory repository. As an example, the *FIPARequest* pattern introduces one possible solution to the recurrent problem to create a conversation among two agents according to the FIPA Request agent interaction protocol.

The structure of the two agents involved in the communication is described by two SASD diagrams (Figure 5.18), which illustrate the attributes and methods to be added to the initiator and participant agents when the pattern is applied to them. A plethora of methods are specifically related to protocol communicative acts; these methods have the preamble "handle" followed by the name of the communicative act. For example, *handleAgree* or *handleInform* is the method in which messages containing the *Agree* or *Inform* performatives will be managed.

These two diagrams do not suffice to describe all the features of the FIPA protocol management because they do not provide any dynamic representation. An MABD diagram is therefore needed to complete the pattern description. This is useful to describe the activities performed by the two agents involved in the communication (Figure 5.19) in a form that can be easily reused as a portion of the actual design of the system (in fact, once a pattern is applied to the project, PTK automatically introduces it in the corresponding diagrams).

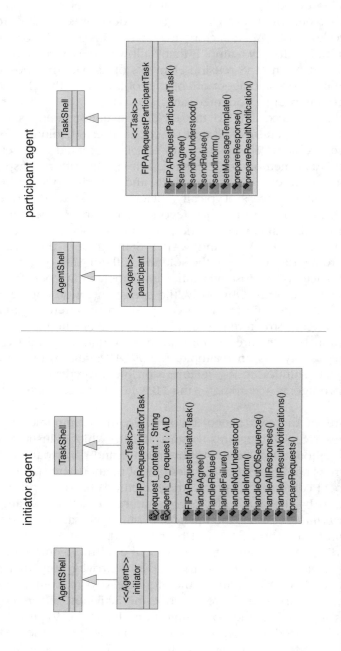

Figure 5.18 Two Class Diagrams Representing Static Structure of Agents Involved in FIPA Request Communication

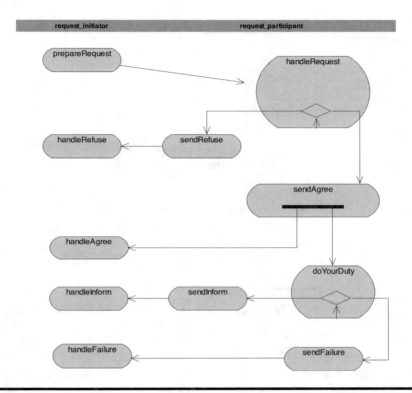

Figure 5.19 A Multiagent Behavior Description Diagram Used to Describe FIPA Request Pattern

The MABD illustrated in Figure 5.19 illustrates that the *request_initiator* agent sends a message to the *request_participant* agent with the *prepareRequest* method (see also Figure 5.18). The responding agent receives it with the *handleRequest* method and, according to its will, responds with a message containing one of the "request" interaction protocol performatives (*Refuse, Agree,…*) sent by the correspondent method (*sendRefuse, sendAgree,…*).

Because a significant part of the design and an even more substantial part of the code automatically are contingent on the appropriate choice of the right pattern for a specific situation, this activity becomes a strategic one and should not be neglected by the designer.

The Deployment Model

The deployment model is the response to the need to detail the position of the agents in a distributed system or in mobile-agent contexts. The deployment configuration diagram (Figure 5.20) is useful to depict where

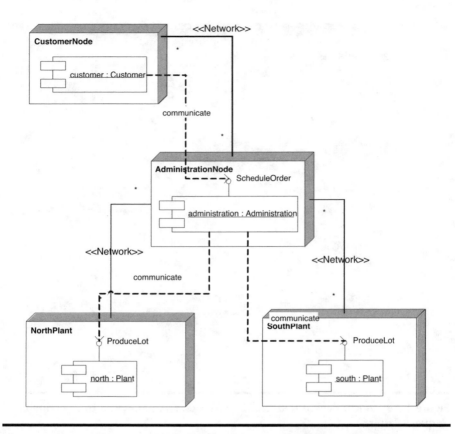

Figure 5.20 Deployment Configuration Diagram for Agents of PS-Bikes' System

the agents will be located during their life (i.e., the processing units where they live), their movement, and their communication support. The standard UML notation is useful for representing the elaborating units, here shown as three-dimensional boxes, and the agents, which are depicted as components; because an agent may be instantiated more than once, agent (instance) names are in the form *agent-name:agent-class*.

Agent and Society Test

The testing activity in PASSI has been split into two different steps: the (single) agent test and the society test. During the agent test, the aim is to verify whether each single agent respects its specifications because these can be derived from the different design steps. Most test cases can be derived from the use cases that constitute the agent functionality specification as described in the agent Identification diagram.

In the society test, the validation of the correct interaction of the agents is performed to ascertain that they concur in solving problems requiring cooperation. Only at this stage is it possible to verify whether the expected social behavior is achieved and the agents interoperate correctly without any problems. This is also the moment for evaluating the system performance in terms of:

- Results provided by the different agents making it up (i.e., if they are able to offer the required services or to deal with the required amount of data) while interacting with the others in the real operating configuration.
- Effect that the operating environment (network capabilities, host platform's elaboration power, and configurations) has on the system.

AGENT IMPLEMENTATION

A distinguishing feature of the PASSI methodology is that it covers the whole development process from requirements analysis to code implementation. The aim of this section is to conclude the overview of the agent modeling process with a concrete realization of an agent, starting with the requirements analysis (system requirements model) and continuing up to the social representations (agent society model) of the agents involved and their architectural implementation details (agent implementation and deployment models).

In this section, a brief description of the programming code derived naturally from the process diagrams will be given. Part of this code has been generated by PTK, and only a few lines have been added manually. The solution presented is an implementation in JADE of a portion of the *Administration* agent already described in the previous design phases (namely, the part dealing with the main agent class and a behavior that initiates a "request" communication).

Figure 5.21 shows a portion of the code for the *Customer* agent. The internal structure of the *RequestWork* behavior has by now been omitted (at lines 30 to 62) because in a first phase we focus on another issue, that is, the agent inner structure represented by its base class.

Line 1 is the declaration of the *Customer* agent as a Java class inherited by the JADE *Agent* class (i.e., the mother class of all JADE agents; see also the section on the JADE implementation platform). Line 2 defines an agent attribute, called *AGENT_NAME*; there is no difference between an agent attribute and a class attribute because both follow the same (Java) syntax. This attribute (a string constant) has been introduced in the agent to contain its name; this value may be used in order to register some agent services to the local directory facilitator (DF).

```
1    public class Customer extends Agent {
2        private final String AGENT_NAME = "customer" ;
3        private Order order;
4        private CustomerBook book;
5
6        public Customer() {
7            initialize_customer_book();
8        }
9
10           public void setup ( ) {
11               register_to_df();
12           GUIManagerTask gui = new GUIManagerTask(this);
13           addBehavior(gui);
14           }
15
16           public void register_to_df ( ) {
17                   /* this block enables DF registration */
18                   try {
19                           // create the agent description of itself
20                           DFAgentDescription dfd = new DFAgentDescription();
21                           dfd.setName(getAID());
22
23                           // register the description with the DF
24                           DFService.register(this, dfd);
25                   } catch (FIPAException e) {
26                           e.printStackTrace();
27                   }
28           }
29
30           public class RequestWork extends AchieveREInitiator {
...              ...
62           }
63
...      ...
268  }
```

Figure 5.21   Portion of Code for the *Customer* Agent Base Class

The constructor method (Customer) in this case is used to call another method in which the customer book is initialized (not dealt with by our example). Agents' constructors are often used only to initialize data structures, while the agent behavior is delegated to the methods that follow.

The agent *setup* method is declared at line 10. An agent may contain several methods, but some of them are reserved for specific goals. The *setup* method is one of them and is a mandatory element because it represents the starting point for all agent activities. Once an agent instance has been created (and its base class constructor executed), the platform registers it automatically to the local agent management service (AMS); it then invokes the agent *setup* method.

In our example, the *Customer* agent *setup* method contains only three instructions:

- (Line 11): an invocation to the *register_to_df* method defined some lines later (lines 16 to 28). This method inserts a new record in the local DF register. The instruction used for this operation is at line 24 (*DFService.register(this, dfd)*); it is put inside a *try–catch* construct to intercept possible exceptions arising during the registration. The *dfd* parameter is a *DFAgentDescription* object and represents the record used to describe the agent to the community. At line 21, this record is initialized with the agent ID value.
- (Lines 12, 13) the *GUIManagerTask* behavior is created and then scheduled with the *addbehavior* instruction. This is the classic way to start a new agent behavior in JADE. As can be deduced from the MABD diagram illustrated in Figure 5.17, this behavior will interact with the user and then call another behavior (the *RequestWork* behavior), which is described more in detail below.

Now we can analyze the structure of the *RequestWork* behavior, which was omitted in Figure 5.21 (lines 30 to 62); the complete code is illustrated in Figure 5.22. This is not the only behavior of the *Customer* agent (see also the agent structure described in Figure 5.15 and Figure 5.16), but it has been chosen because it is a classic communication task. The behavior is declared as a *Customer* agent inner class, and it inherits a JADE core superclass whose name is not univocally defined (as it was for the JADE *Agent* class used to define the agent); in fact, a complex hierarchy of behavior types is provided by this implementation platform and the choice is left to the developer. Each element of the hierarchy has its specific functionalities; for example, the *CyclicBehavior* may be used to create a behavior that cyclically repeats an operation, the *SequentialBehavior* to execute some activities in the specified order, and the *FSMBehavior* to implement a complex finite state machine.

The *RequestWork* behavior starts a "request" conversation with the purpose to obtain some service from the *Administration* agent. The JADE API offers an off-the-shelf behavior to initiate a communication by adopting several communication protocols, the *AchieveREInitiator*, and the *AchieveREResponder* to deal with the consequent incoming messages.

Line 30 defines the behavior as a class (*RequestWork*) that extends the *AchieveREInitiator* superclass. It also has some attributes defined at lines 31 to 33: *request_content* is a string containing the message content (coded in a specific content language, e.g., RDF) for the initial "request" communicative act. The other attribute, *agent_to_request* (used to address the receiver agent), is an instance of the AID class belonging to the JADE API

```
30  public class RequestWork extends AchieveREInitiator {
31      private String request_content ;
32      private AID agent_to_request ;
33      private GUIManagerTask gui;
34
35      public RequestWork( Agent owner, AID id, String content, GUIManagerTask gui) {
36          super(owner, new ACLMessage(ACLMessage.REQUEST) );
37          agent_to_request = id;
38          request_content = some_service;
39      }
40      public void handleAgree ( ACLMessage msg ) {
41          gui.notifyOrderAccepted();
42      }
43      public void handleRefuse ( ACLMessage msg ) {
44          gui.notifyOrderRefused();
45      }
46      public void handleInform ( ACLMessage msg ) {
47          gui.notifyOrderScheduled();
48      }
49      public Vector prepareRequests ( ACLMessage msg ) {
50          //automatically invoked by the platform after the class constructor
51          msg.setPerformative(ACLMessage.REQUEST);
52          msg.setProtocol( FIPANames.InteractionProtocol.FIPA_REQUEST );
53          msg.setSender(myAgent.getAID());
54          msg.addReceiver(agent_to_request);
55          msg.setContent(request_content);
56
57          Vector l = new Vector();
58          l.addElement(msg);
59          return l;
60      }
61  }
62
```

Figure 5.22   Portion of Code for *RequestWork* Behavior of *Customer* Agent

framework; this is a container for the univocal identifier used to locate an agent within a specific platform. The *gui* attribute is used to store a reference to the behavior that calls this (*GUIManagerTask*; see Figure 5.19) in order to notify it with the results of the communication.

The *RequestWork* constructor is defined at lines 35 to 39. It requires four parameters: the *owner* (a reference to the agent); the *AID* (the receiver agent's unique ID); the *request_content* (the content of the message to be sent); and the *gui* reference to the caller behavior (see above). The first command of this method is a call to the super class constructor that is invoked by specifying, with the first parameter, the owner agent and, with the second parameter, that the message to be used to initiate the protocol is a *request* communicative act. This last parameter is not of paramount importance because the request message is better defined in the following *prepareRequest* method.

Once the constructor is completed, the *prepareRequest* method (lines 49 to 60) is automatically invoked for all the *AchieveREIntiator* type behaviors. It returns a vector of *ACLMessage* objects used to initiate the communications with *n* different agents. The *ACLMessage* class represents the data structure used to contain the message payload of a message (in ACL language). In this method, the *performative, protocol, sender, receiver,* and *content* fields of the message are filled in with necessary data. Then, at lines 57 to 59 the vector *l* is filled in with the message and the method terminates by returning this vector as a result. At this point, the *AchieveREIntiator* superclass actually sends the message to the receiver agent.

Lines 40 to 48 show the definitions of methods devoted to handling the incoming messages sent by the receiver agent during this communication. It is possible to observe a *handleX* method for each expected communicative act, where the *X* is the name of the performative (*inform, agree,* ...). In this way, when an *agree* message reaches the agent, the *handleAgree* method is invoked with this message as a parameter.

What can be derived from the code described in this section is that coding FIPA agents under the JADE platform is essentially JAVA coding. The most important difference is not in the actual agent code, but in the communication infrastructure offered by the platform that acts like a middleware, enabling agents of our system to interact easily and relieving the designer of many decisions regarding details. For instance, the designer does not need to know the location of a mobile agent at a given moment to code a message for it; the simple agent unique name is sufficient, and the AP will then take care of correctly delivering the message. This, in essence, is the mission of FIPA: to enable the interoperability of heterogeneous software agents.

REFERENCES

1. Searle, J.R., Speech Acts, Cambridge University Press, 1969.
2. Cranefield, S. and Purvis, M., UML as an ontology modeling language, in Proc. Workshop Intelligent Inf. Integration, IJCAI-99, Stockholm, Sweden, July 1999.
3. Bergenti, F. and Poggi, A., Exploiting UML in the design of multi-agent systems, in Engineering Societies in the Agents World — Lecture Notes on Artificial Intelligence, Omicidi, A., Tolksdorf, R., and Zambonelli, F., Eds., Springer Verlag, Berlin, 1972, 106, 2000.
4. Cossentino, M., Different perspectives in designing multi-agent systems, in Proc. AGES '02 Workshop at NODe02, 8–9 October 2002, Erfurt, Germany, 2002.
5. Cossentino M. and Potts, C., A CASE tool supported methodology for the design of multi-agent systems, in Proc. 2002 Int. Conf. Software Eng. Res. Practice (SERP'02), June 24–27, 2002, Las Vegas, NV, 2002.
6. Bauer, B., Müller, J.P., and Odell, J., Agent UML: a formalism for specifying multiagent interaction, in Agent-Oriented Software Engineering, Ciancarini P. and Wooldridge, M., Eds., Springer, Berlin, 91, 2001.

7. Parunak, H.V.D. and Odell, J., Representing social structures in UML, in Proc. Agent-Oriented Software Eng. (AOSE) — Agents 2001, Montreal, Canada, 17, 2001.

8. Bellifemine, F., Poggi, A., and Rimassa, G., JADE — a FIPA2000-compliant agent development environment, in Proc. Agents 5th Int. Conf. Autonomous Agents (Agents 2001), 216, Montreal, Canada, 2001.

9. FIPA request interaction protocol specification, FIPA document no. 00026, 2002, available at http://www.fipa.org/specs/fipa00026/.

10. Rumbaugh, J., Jacobson, I., and Booch, G., The Unified Modeling Language Reference Manual, Addison–Wesley, Reading, MA, 1999.

11. OMG unified modeling language specification — version 1.5, OMG document formal/03-03-01, March 2003.

12. FIPA methodology glossary, available at http://www.pa.icar.cnr.it/~cossentino/FIPAmeth/glossary.htm.

13. Carlson, D., Modeling XML Applications with UML, Addison–Wesley, Reading, MA, 2001.

14. Resource description framework — (RDF) model and syntax specification, W3C recommendation, 22-02-1999, available at http://www.w3.org/TR/1999/REC-rdf-syntax-19990222/.

15. FIPA RDF content language specification, Foundation for Intelligent Physical Agents, document FIPA XC00011B (2001/08/10), available at http://www.fipa.org/specs/fipa00011/XC00011B.html.

16. Odell, J., Parunak, H.V.D., and Bauer, B., Representing agent interaction protocols in UML, in Agent-Oriented Software Engineering, Ciancarini P. and Wooldridge, M., Eds., Springer Verlag, Berlin, 121, 2001.

17. Poslad, S., Buckle, P., and Hadingham, R., The FIPA-OS agent platform: open source for open standards, in Proc. 5th Int. Conf. Exhibition Practical Application Intelligent Agents Multi-Agents, Manchester, U.K., April 2000, 355–368, 2000.

18. Gamma, E., Helm, R., Johnson, R., and Vlissides, J., Design Patterns: Elements of Reusable Object-Oriented Software, Addison–Wesley, Reading, MA, 1995.

19. Cossentino, M., Sabatucci, L., and Chella, A., A possible approach to the development of robotic multi-agent systems, in Proc. IEEE/WIC Conf. Intelligent Agent Technol. (IAT'03), Halifax (Canada), 2003.

20. Cossentino, M., Sabatucci, L., Sorace, S., and Chella, A., Pattern reuse in the PASSI methodology, in Proc. ESAW'03 Workshop, Imperial College, London, U.K., 2003.

6

PAST SUCCESSES

This chapter details several of the most outstanding applications of multiagent systems (MASs) designed to provide manufacturing systems with agility. In particular, to keep the review in line with Chapter 3, the successful stories presented here have been organized into the following sections. After a brief introduction in the first section, the next section outlines the details of two comprehensive models that provide a supply chain-oriented solution to agile planning and scheduling in manufacturing. The third and fourth sections are devoted to MAS applications to planning and scheduling (P&S) and to scheduling and control (S&C) in manufacturing, respectively. In addition, the fifth section presents some of the most prominent successful MAS applications found in industry. The chapter closes with comments the previous discussion and draws conclusions.

INTRODUCTION

Before reviewing the current success stories of agent-based manufacturing, it is necessary to return for a moment to modern manufacturing enterprises' need for agility and to the identification of the characteristics of an information framework aiming to make manufacturing agile. Agile manufacturing was broadly defined in the first chapter, and the following definition of Cho et al. [1] can effectively summarize the concept in a few words: "agile manufacturing is the capability of surviving and prospering in a competitive environment of continuous and unpredictable change by reacting quickly and effectively to changing markets, driven by customer-designed products and services."

According to Gunasekaran [2], a conceptual model for the development of agile manufacturing systems (AMSs) should be developed along four key dimensions: strategies; technology; systems; and people. Regarding strategies, some of the strategic decision areas that should be taken into account while developing AMSs are concurrent engineering (viewed as a more systematic method of concurrently designing the product and the downstream processes for production and support); rapid partnership formation; strategic alliances; virtual enterprise; and physically distributed manufacturing systems [3]. Technology can enable a rapid hardware changeover by the use of robots; flexible part feeders; flexible fixturing; modular grippers; and modular assembly hardware; equipped with flexible software technologies such as Internet and multimedia technologies. Systems should satisfy the development and integration of information infrastructure to facilitate distributed design, planning, manufacturing, and marketing functions in the agile enterprise; in the meantime, however, the integration of current fragmented computer systems is perhaps the biggest challenge faced by the AM enterprise [2]. Finally, agility demands close collaboration among all personnel and team members, information technologies, such as collaborative and workflow systems, by themselves do not suffice to achieve the desired communications efficiency. Strategies to overcome these factors, e.g., team training and project planning, should therefore be adopted in the transformation to agility.

The discussion in the previous chapters should have placed in evidence the important role that agent- and/or holonic-based solutions can play in enabling agility in manufacturing. MASs aptly represent a mixed set of strategies, technologies, and systems that can be used to enhance or to add agility. Whatever the details of the specific implementation adopted are, MASs can enable at least two of the main behaviors required for an MS to be agile: to distribute decisional and operational capabilities among its components and to be reactive and adaptable to the changes observed in the environment.

All this said, it is up to people to use these tools. Thus, agent- and/or holonic-based solutions can be seriously considered as the right answer for agility, especially for small medium manufacturing enterprises. This chapter sets out to illustrate many possible applications of MASs, in particular focusing on the planning, scheduling, and control aspects. This review, in fact, seeks to provide a valuable reference tool for researchers and practitioners: for the former, to settle the issue about what has been done in the field, and for the latter, to evaluate the possible adaptability and appropriateness of such solutions to the requirements of their current MS.

TWO OUTSTANDING SUPPLY CHAIN-ORIENTED MAS APPLICATIONS IN MANUFACTURING

This section reviews in detail two of the most outstanding approaches of MASs to manufacturing that were already mentioned in Chapter 3. Attention is focused here on the Metamorph II [4] and the AARIA [5] projects, which epitomize different MAS architectures introduced to provide agent-based agility to manufacturing systems. In particular, it is worthwhile to consider the enlarged view of both the applications to embrace the whole supply chain where an MS is included and the way in which this is achieved with different MAS models.

The Metamorph II Project

The Metamorph II project has been the subject of a number of papers by members of the research team of the Division of Manufacturing Engineering at the University of Calgary (the history and main characteristics of the project are summarized in Shen and Norrie [4]). This project is the result of different earlier projects by the Calgary team that aimed to exploit multiagent architectures to support in an intelligent and integrated way the several activities that take place in MS. In particular, a first project called Distributed Intelligent Design Environment (DIDE) [5] introduced intelligent agents to set up a network for interconnecting design and engineering specialists, represented by engineering tools, CAD/CAM tools, and knowledge-based and database systems. The project produced an open system in which components and competence can be dynamically added and used by activating an associated agent. Thus, DIDE is a striking application of an MAS to coordinate design capabilities, as well as to share and manage knowledge in MS.

The second precursor was an agent-based system for intelligent manufacturing, the Metamorph I project [6]. Agents were associated with manufacturing devices, products, or parts, and mediator agents existed to coordinate their interactions. The project's purpose was thus to establish a distributed agent architecture that enabled agent collaboration in solving problems (specifically relevant to P&S in a single MS) using an approach based on problem decomposition and on distributing decision responsibilities.

Metamorph II has evolved from its predecessors, extending their approaches to integrate all the possible activities characterizing an MS in a supply chain into a distributed intelligent system: from internal activities relevant to design, planning, scheduling, and operational execution, to external activities such as procurement and supply order management, to distribution and customer order management. The architecture adopted by Metamorph II is a federation based on mediator agents. The functions, which are associated with the production cycle of a manufacturing enterprise (viewed here as the center of a network composed of other enterprises and customers, which constitutes the enterprise's supply chain), are distributed in a network of agents with different roles. The mediators coordinate the activity of subsystems that, in turn, may include other agent-based or traditional legacy systems.

The elements making up the higher level of the Metamorph II architecture correspond to single or sets of interconnected mediators, whose activities are relevant to different functional areas, namely, the partners, marketing, design, planning and scheduling, execution and material supply areas. For example, the material supply is managed by a Material Mediator which is connected to the company Enterprise Mediator operating in the planning and scheduling area, to one or more Enterprise Mediators populating the partner area, to one or more Design Mediator in the design area, and to the Marketing Mediator in the marketing area in order to enable the coordination of the material supply activity with the company production objectives.

The central role in the Metamorph II architecture is handled by the Enterprise Mediator, which registers the presence of any other mediators in the system (i.e., in the supply chain) and coordinates the enterprise activity at the higher level: in particular, such a mediator exchanges information with other mediator agents that are representative of the same level of aggregation, namely, other Enterprise Mediators; these latter are the gateway through which the procurement and supply processes take place, that is, through which the requests and orders between the enterprise and its customers and suppliers are communicated. The high level functional area devoted to planning and scheduling is composed by an Enterprise Mediator and a set of Resource Mediators. The whole planning and scheduling area is implemented by several hierarchies of mediator agents which are rooted by a Resource Mediator; this kind of agent is associated with a shop floor or part of it, and is responsible for distributing and managing its assigned production activities. A Resource Mediator coordinates, in turn, a number of lower level agents, which correspond to Machine Mediators, Tool Mediators, Transportation Mediators, and Worker Mediators, that are associated with a specific class of resources available in the shop floor. Finally, each of these mediators coordinates the activity of the agents associated with the single resources of a specific class, e.g., machine agents or tool agents.

The organization of agents in Metamorph II does not reflect a bureaucracy as it does in a hierarchical architecture; on the contrary, the mediators introduced at the different levels are used to distribute the decision complexity among homogeneous subsets of actors and to adopt a negotiation-based approach to select the best alternative. Mediators neither impose a predefined plan or activity to their lower level agents nor centrally organize the schedule of the resource activity by means of some dispatching rule or algorithm; in fact, they call for bids for the execution of an activity, successively collect the replies, and then select the best offer on the basis of a performance criterion. In this way, a plan and a schedule are defined by a combination of a mediation and a bidding mechanism.

For example, a request from a customer is forwarded by the marketing mediator to a resource mediator (directly or through the enterprise mediator) in order to start a planning cycle to determine cost and delivery time of a possible order. The resource mediator then asks for bids from the lower level mediators by sending a *request* message to machine or worker mediators, which coordinate the shop floor resources needed to manufacture that order. Actually, such a call can be achieved with a machine-centered or a worker-centered strategy, depending on which of these mediators has priority in the P&S problem decomposition (here, machines are assumed to have priority). The machine mediator asks (with an *announce* message) each available and appropriate machine agent for a bid; note that the bid must include the information on cost, time, and resources to be used. The machine agents, in turn, send a *request* for a bid to the tool mediators and worker mediators needed for that activity, and those mediators forward an *announce* message to the tool and worker agents, respectively.

The replies (*bid* messages) from the single resource agents are collected by the pertinent mediator, which selects the best one (e.g., on a cost basis) to which to reply and communicates this along the hierarchy of the involved mediators (with an *inform* message) up to the enterprise mediator. This mediator finally selects the best overall bid, corresponding to the best possible plan for the order. This is the defined plan, which is proposed to the customer through the original marketing mediator. If the proposal is accepted, a sequence of *award* messages is forwarded again from the mediators to the single resource agents that offered the best bids in order to acknowledge the acceptance of the proposed plan; this plan must be registered by the single resources (e.g., machines, tools, workers), thus becoming part of their schedules.

Figure 6.1 depicts the exchange of messages among the various agents involved, starting from a Resource Mediator request and focusing in particular on machine agent 1 that here is assumed to be the one ultimately selected to serve the request (in the figure boxes represent mediators whereas ovals specific agents).

The details of the distributed bidding-based scheduling mechanism adopted in Metamorph II are provided in [7]. This mechanism uses an asynchronous communication scheme among the agents with two classes of messages. *Request and assertion* messages in particular are:

- *Request*: to ask for a kind of service
- *Inform*: to report the selected (as well as alternative) resources
- *Notice*, to notify whether a bid has been selected and the task has been assigned

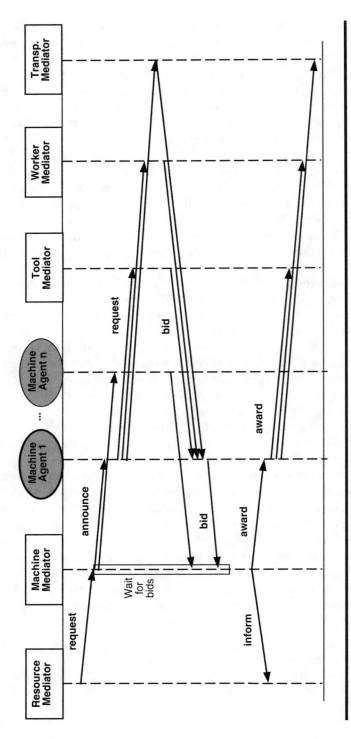

Figure 6.1   The (Partial) Exchange of Messages among Metamorph II Agents Processing a Request

The second class, *call for bids and offers* messages, is:

- *Announce*: to ask for a bid to provide a service
- *Bid*: to reply to an announce message

The shop floor resource mediator initiates the bidding process by selecting from the list of incoming orders the first ones to be considered for concurrent manufacture in order to satisfy a predefined production mix. The cost model used to evaluate the bids is obtained as the combination of processing, set-up, and fixed costs paid for the use of the machines, tooling, and subcontracting costs, plus penalty costs for the possible late completion of the orders. The interaction among the agents basically follows the auction schema of the contract net protocol. Note that the proposals of resource agents may conflict because the same time slots can be offered to execute different tasks. During the commitment phase, the shop floor mediator resolves possible conflicts by assigning a priority to the conflicting tasks on the basis of their cost. Whenever a task with a lower priority cannot be executed because the proposed time slot becomes unavailable, the task is returned at the top of the order request list to be considered again in the next planning phase.

A more recent project related to Metamorph II developed at the University of Calgary by the Intelligent Manufacturing System Group (IMSG) and the Knowledge Systems Institute (KSI) involves a Collaborative Agent System Architecture (CASA) project and an Infrastructure for Collaborative Agent Systems (ICAS) project, which is also used for P&S in MS [8]. The project's purpose is to define a framework and an architecture for the development of MASs through the customization of a set of predefined domain-independent services. The project extends and generalizes the concepts introduced in the Metamorph II system, giving particular attention to the infrastructure used by the agents to communicate, share data, and register or locate services. P&S can be modeled again by adopting a bidding-based mechanism that is preceded by a collaboration phase during which knowledge-based agents are queried to locate the appropriate resources (i.e., agents) for the required operations.

The AARIA Project

The Autonomous Agents for Rock Island Arsenal (AARIA) project [9] was developed by the research team of ERIM-CEC (Centre for Electronic Commerce, Ann Arbor, Michigan) to tackle and, to a certain extent, control, the P&S for an army facility. Specifically, the AARIA agent-based architecture was designed to provide a solution to industrial requirements, and its developers did not consider it only as a research product. This overriding industrial approach and the ensuing discussion about the requirements and features exploitable in a manufacturing environment placed the AARIA project in the spotlight. In fact, its authors compare the agent solution of AARIA with other nonagent software in order to show how its advantages make it more suitable for an MS.

The requirements that the AARIA architecture aims to satisfy have been outlined in Parunak et al. [9] and are briefly reviewed here:

- Shop floor scheduling and control: in particular, this point amounts to the provision of a P&S system with the ability of dynamic scheduling and rescheduling.
- External interfaces: the system must be strongly integrated with the supply chain of the MS, supporting the relationships with customers and suppliers. In addition, the system must allow the interface with the personnel inside the MS, such as operators, who are ultimately responsible for the decisions about the production process, and manufacturing engineers and managers, who make decisions about product design and maintenance, modifying the availability and the characteristics of the production resources.
- Internal operations: the system must have functions and capabilities to manage the business processes in the "inner side" of the supply chain. Particularly, again, it should be able to plan, covering the classic ERP functionalities, but also to cope with the frequent changes that occur in the market (new and changed orders) and in design (new products and customizations). It must likewise be capable of performing dynamic scheduling and control of the execution of operations. Two aspects are emphasized: metamorphosis and uniform interface. The first feature highlights the need for a system composed of not only static entities, but also entities that change; they are generated, transformed, and, possibly, eliminated according to the evolution of the physical (or logical) items with which they are associated. A uniform interface means being able to provide the entities in any part of the system with a common interaction protocol and ontology, so that external relations (with customers and suppliers) and internal (with other entities in the MS) relations take place uniformly. This latter feature is pivotal for gradually introducing new software for managing an MS (or an MS subsystem) and for smoothly scaling it to extend to the whole enterprise.

AARIA's architecture is built on a philosophy whose guiding principle is to populate the system with a number of agents that monitor the environment, act autonomously, and, possibly, are born, live, and die.

Agents are identified by adopting a design methodology that basically translates a declarative description of a system, which specifies items and the events involving items of interest (details about this are given in Parunak et al. [10]). Agents then derive from a combination of the physical and functional decomposition of the MS of concern: persistent agents are associated with parts, resources, and unit processes and transient agents are introduced to model the interactions among persistent agents.

A part agent (or part broker) corresponds to a class of parts or components manufactured in the system that store state information about the class, such as availability in the inventory, production history, forecasts, and so on. The resource agents (or resource brokers) are associated with individual shop floor resources, e.g., machines, tools, or human operators, and store information about their schedules. The unit process agent (or unit process broker) corresponds to a function — that is, it possesses the knowledge about the process needed to produce some result, e.g., which parts and resources are needed to manufacture a product. This type of agent is then associated with workflows in the MS. Transient agents, like engagements, materials, and products, model the interactions between a unit process and other types of agents, whereas operation agents model the evolution of a unit process. The existence of transient agents follows a cycle of six states (inquiring, committing, committed, available, active, and archived) during which the agents are responsible for the activities of a specific process and involving determined resources and physical parts. As an example, one can consider the way in which a customer request about a product is processed (as illustrated in Figure 6.2, redrawn from Parunak et al. [9]).

The customer is associated with a unit process agent and the request generates a transient material agent, which includes information about the desired product, quantity, and delivery time. The material agent is initially in the *inquiring* state; it replies to the customer (i.e., the associated agent) with a worst-case bid (including price and delivery) and then starts to analyze the request and possibly to reduce the bid. The material agent queries the persistent part broker about the desired product; the bid is updated depending on the available inventory for the product or on the need to start a production planning cycle for it. Whenever the customer agent accepts a bid, it changes its state to *committing* and then to *committed* after the material agent (which, in turn, has changed its state to *committed*) confirms the agreement. If the broker agent confirms the inventory availability of the product, the material agent moves directly to the *available* state.

Conversely, a production cycle is initiated to satisfy the request, and the material agent propagates the customer commitment throughout the chain of the other transient agents involved in the production, e.g., the

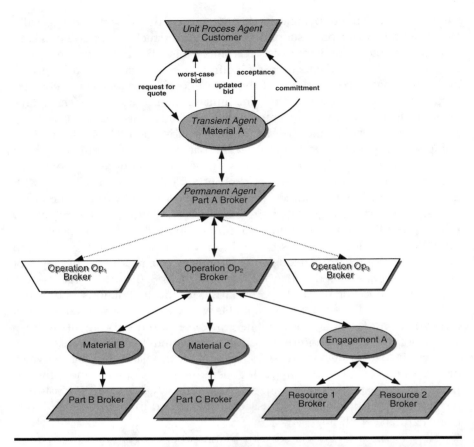

Figure 6.2   Example of Customer Request for Product Processed by AARIA Agents (Redrawn from Parunak, H.V.D. et al., *Integrated Computer-Aided Eng.*, 8, 45, 2001)

product agent, the subcomponent material agents, and the engagement agent. In turn, these change their state from *committed* to *active* (e.g., when the production operations are in execution), then to *available* (e.g., when the product has been manufactured), and finally to *archived* (e.g., when the transient agents complete their task and the relevant production history is stored in the persistent agents). The revenue obtained from the customer for the successful supply is distributed among the cost centers that have been involved. Some points are worthy of comment:

■ Only persistent agents can start an operation.
■ The schedule derived from resource engagement is kept flexible so that it can be revised and changed according to variations in customer

requests, possible failures, or optimization of production (e.g., clustering similar orders into a larger lot to reduce the set-up cost).

■ Different types of persistent agents may initiate an operation according to different criteria, thereby adapting the MS behavior to different modalities. This feature, denoted as *modality emergence*, allows switching from a push to a pull production modality. A unit process agent can start a production activity according to a predefined schedule, a part agent can do the same whenever its inventory level drops below a fixed threshold, or a resource agent can call for an operation whenever it becomes available.

The resulting system behavior is thus highly flexible; the system's evolution is not fixed, but the agents, exploiting their autonomy, generate a system whose behavior is not predefined and adapt it to circumstances through the provision of flexibility. The uniformity feature makes it possible for even the components of the single MS to interrelate as though they constituted a sort of supply chain: the agents interact in the internal supply chain (with internal suppliers and customers) as they do in the external supply chain (with the actual company customers and suppliers). Thus, the AARIA architecture was designed to scale from a single MS to support the business process of a network of interconnected MSs. This characteristic seems to guarantee a fitting solution for enabling the Internet connectivity of an enterprise, from e-commerce to e-business and e-procurement.

SUCCESSFUL MAS APPLICATIONS IN MANUFACTURING PLANNING AND SCHEDULING

This section reviews several relevant approaches to exploit MASs in agile manufacturing, focusing in particular on planning and scheduling (P&S) applications. Following the classification introduced in Chapter 3, the applications considered here can be viewed as representative of the various approaches following the schema of Table 6.1.

Auction-Based Models

The system proposed by Gu et al. [11] is representative of a bidding-based model for P&S, which shares some similarity with the Metamorph II and the AARIA approaches. In fact, these authors associate the autonomous agents with machines, tools, parts, and the shop manager. In particular, a part agent represents a product order; such an agent knows the functional requirements of the pertinent product, which are represented by the standard for exchange of products (STEP) data model of the Interna-

Table 6.1   Possible Classification of Reviewed Agent-Based P&S Applications

Classes of agent-based P&S models	Reviewed applications
Auction based	Gu et al. [11]
	HOLOS, Rabelo et al. [12, 13]
	Sousa and Ramos [15]
	Liu and Sycara [16]
	Archimede and Coudert [17]
Cooperation based	Frankovic and Dang [18]
	Kouiss et al. [19]
	Sikora and Shaw [20]
Hierachical	ProPlanT, Marik et al. [21]

tional Organization for Standardization (ISO); by the batch size for the order; and by its due date. However, the proposed MAS differs from a pure autonomous agent approach in that a hierarchy organization is introduced among the machine agents in order to regulate communications and to take into account the implicit hierarchy imposed on the machines by the order of precedence among the operations needed to manufacture the products.

Assuming the production of homogeneous products, a hierarchical organization can be forced on the machine agents depending on the kind of operation the machines are able to perform. In order to satisfy an order, the shop manager agent broadcasts a request to the machine agents at the highest level; these start a planning process by determining whether the machine can provide each feature of the part to manufacture. If the reply is affirmative, the machine agents ask the tool agents to select the appropriate tools for the operation; otherwise, they subcontract the operation to the machine agents at the next level. During this hierarchical examination of the task features, the machine agents also verify whether the tolerance requirements for the part can be met and locally optimize the set-up sequence they must execute in order to reduce the number of tool changes.

Planning in this case is also an "ask-for-bid" process that starts from a central coordinator, the shop floor manager, and is actually performed by the machine agents following the priority imposed by the agent hierarchy. Higher level agents receive replies for bids from lower level agents. Scheduling is again the propagation of the commitment associated with the selection of the best bid from higher to lower levels.

It is immediately evident that the central role yielded by this approach to the machine agents suggests its appropriateness for quite structured production environments and for complex multifunctional flexible manufacturing cells.

The MAS application by Rabelo et al. [12] reflects the outlook on scheduling in manufacturing, which has shifted from optimality to flexibility and finally to agility. Agile scheduling denotes the ability to react dynamically to unforeseen events (process flexibility) and to take into account all of the production resources of a "virtual" enterprise, rather than only the ones present at the shop floor level (boundary flexibility). These authors report their experience with the HOLOS multiagent framework within the Multiagent Agile Manufacturing Scheduling Systems for Virtual Enterprises (MASSYVE) project sponsored by the EU. They sought to extend the HOLOS framework, initially developed to support the generation of individual agile scheduling systems, to define scheduling applications at a virtual enterprise level.

A multiagent scheduling system is viewed as a network of heterogeneous processors (the agents) that are associated with manufacturing resources and able to perform autonomous decision-making activities on the basis of their knowledge and to communicate and cooperate with each other. The HOLOS generic architecture is based on the model proposed by Rabelo and Camarinha–Matos [13]; because functional and physical agents are introduced, it is a hybrid architecture and has been defined by the authors as *semihierarchical* given its layered organization and roles of agents. Three classes of permanent agents are included: the *scheduling supervisor agent* (SSA) responsible for invoking a global scheduling activity; *enterprise activity agents* (EAA) associated with the shop floor manufacturing resources; and *local spreading centers* (LSC) acting as broker agents associated with clusters of homogeneously functional EAAs. Also included is a class of transient agents, *consortium* (C), whose purpose is to supervise the execution of the specific order and that remain active as long as the orders are processed.

Figure 6.3 depicts the structure of the MAS by Rabelo et al. The gray-shaded boxes denote agents acting in the planning phase, whereas white boxes indicate agents operating on-line during the schedule execution (note that SSA and EAA are active in both phases). The scheduling activity is based on the contract net protocol and the communication among the agents follows a vertical hierarchy, which derives from the different roles of the agents' classes; in essence, after a task (order) is announced, the LSC agents assign it on the basis of the best bid received from the EAA. We will return to the aspects relevant to on-line control in the section on successful MAS applications in manufacturing scheduling and control.

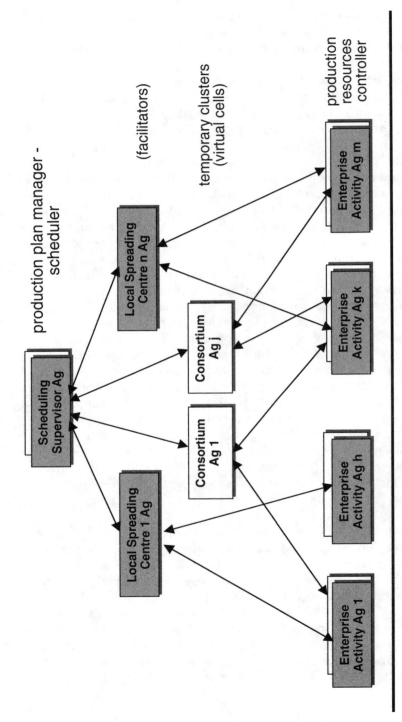

Figure 6.3 MAS Structure for P&S (From Rabelo, R.J. et al., *Robotics Autonomous Syst.*, 27, 15, 1999)

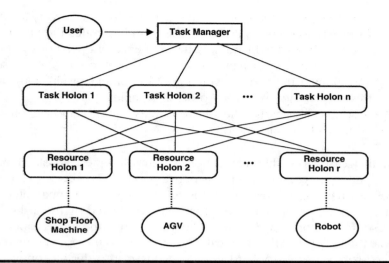

Figure 6.4   Structure of the Scheduling Holon (Redrawn from Sousa, P. and Ramos, C., *Computers Ind.*, 38, 103, 1999)

The HOLOS scheduling system represents the inner level, the so-called intra-organization federation layer, of the three-layered MASSYVE architecture, whose purpose is to provide a "virtual" organization made up of several enterprises or by a multisite enterprise, with an appropriate integration of information that enables cooperation. Planning and scheduling at the various levels of such a network enterprise must access the appropriate subset of updated information from the CIM information system. The solution adopted in MASSYVE is to exploit an object-oriented DBMS, the PEER federated database system [14]. Supporting the sharing and exchange of information among the nodes of the network without the need for central control or data redundancy, the federated database is the backbone for the two upper layers of the MASSYVE architecture: the federation of HOLOS systems, thereby cooperatively managing the manufacturing resources of the different enterprises in the network, and the federation of virtual enterprises that, at the highest level, allows the enterprises to share information in order to cooperate in market opportunities.

Observing how scheduling is basically an activity involving several distributed actors and parallel processes, Sousa and Ramos [15] proposed a holonic scheduling system as a component of an intelligent MS. Holons in a holarchy are quite similar to agents in an MAS, if one disregards the fact that a holon can contain other holons. Thus, the scheduling holon in [15] (see Figure 6.4) can be considered an example of an MAS scheduler made up of three classes of holons (agents):

- The task manager holon, which is the interface with the users, receives the requests for new orders and activates the task holons.
- The task holons, which know the sequence of operations to process and the order, must negotiate with the resources to receive service and monitor the correct evolution of the order's production, and are then deactivated when the order is completed.
- Resource holons handle the resources' agenda and negotiate the request of the task holons.

In this case, too, the negotiation scheme adopted is based on the contract net protocol; however, it is performed in two phases — a forward influence phase and a backward influence phase — to guarantee compliance with the order deadline. After being contacted by the task holons, the resource holons, which are able to execute the order operations, analyze their agenda, and those called on to process the operations without predecessors start the forward influence phase in order to condition the starting times of the downstream operations appropriately so that the sequence is respected. After the forward influence phase is completed, the resource holons in charge of the final operations start the backward influence phase to communicate upstream the actual time constraints for the operations. The final list of time intervals for the operations is then presented to the task holons, which can accept and reserve them or not. The task manager holon has the additional duty of preventing deadlocks or overbookings arising from multiple requests from different task holons for the same resources. The simple strategy used is that of assigning a priority (e.g., first come, first served) to the incoming orders. The task manager keeps track of the resource currently involved in negotiations and activates the task holons that do not have conflicting requests for such resources according to the priority assigned.

MA approaches for manufacturing P&S can be introduced only in specific modules supporting a specific MS decision-making activity. As a matter of fact, MA approaches can also be viewed as heuristic solution paradigms for suitable classes of decision problems, in a similar way to natural or evolutionary methods like simulated annealing, genetic algorithms, or ant colony systems. Scheduling applications in manufacturing, in particular, have attracted the attention of researchers who aimed to exploit two important features of MA approaches, i.e., (1) problem decomposition through the decentralization of decisions to local decision makers; and (2) identification of a global solution as an iterative coordination process consisting of negotiation among the single decision-makers in order to resolve conflicts that may arise.

Liu and Sycara [16] proposed an MAS to face a job shop problem in order to find a "satisfying" feasible solution, i.e., to minimize the weighted tardiness of the jobs. They defined a problem-solving model, called coordinated negotiation agents (CONA), based on two classes of agents: job agents and resource agents. Job agents are in charge of satisfying constraints relevant to the jobs, such as release times, due dates (in case of feasibility problems), and precedence among operations; resource agents handle resource capacity constraints. Coordination through negotiation takes place among job agents and resource agents: the resource agents determine a scheduling proposal by assigning their available capacity to the job operations and the job agents then match the proposal to possible constraint violations and, in case of an unsatisfactory result, propose changes to the tentative schedule. In turn, resource agents check the capacity utilization emerging from the new proposal and resolve possible conflicts by taking into account the job weight and limiting the changes to the schedule of bottleneck (i.e., highly utilized) resources as much as possible.

Job and resource agents use heuristics to formulate counterproposals that should be more easily accepted. Mechanisms based on the maximum number of proposed changes and on tracking of previous capacity assignment are introduced in order to prevent the unbounded cycling of the negotiation process. Even if the convergence of the algorithm is not guaranteed, a trial campaign showed that the CONA approach is able to identify sound solutions quickly — even in case of problems with serious bottlenecks. The optimization of the global cost is sought by assigning to one of the agents associated with a bottleneck resource, heuristically selected, the special role of fixing the subschedule of the operations for that resource by locally optimizing the cost objective and thus controlling in this way the negotiation process. Even in this case, no certainty of finding an optimal solution exists; however, high-quality schedules can be reached in a reasonable computation time.

One can observe Liu and Sycara [16] did not explicitly consider the problem of tackling dynamic or on-line changes by MA scheduling approaches. Archimede and Coudert [17] stressed this aspect in a general MA model, called supervisor, customers, environment, and producers (SCEP) and defined as a framework to develop reactive scheduling applications in manufacturing. The components of the model are in fact two classes of agents: the customer agents, usually associated with manufacturing orders, and producer agents related to the machines. Additionally, a blackboard environment is used to make the agents communicate and to store the system state and, finally, a supervisor agent is responsible for controlling the cooperation cycle. The model is able to cope with workshops characterized by a set of jobs (customer orders) that must be produced by a set of machines, as occurs in a open shop, but with two major differences: no precedence order is prefixed among the jobs' operations, and one or more uniform multifunctional machines are available. The scheduling cost considered is the sum of the cost paid by the jobs for the utilization of the machines.

The MA approach followed in the SCEP model is, in principle, similar to the one in Liu and Sycara [16]; however, in this case, the convergence of the coordination algorithm has been formally demonstrated, and its suitability to realtime situations in which the predefined schedule must be adapted to unpredictable variations (e.g., machine breakdown) has been taken into account. The drawback of the SCEP approach is traceable to the neglect of set-up and transportation times, despite the fact that the flexible workshop model considered, which includes uniform multifunctional machines and no prefixed order (i.e., routing) of the jobs' operations, should require them.

Cooperation-Based Models

The system conceived by Frankovic and Dang [18] can be considered an important MAS model following the cooperation-based approach discussed in the subsection concerning planning and scheduling MAS approaches based on a cooperation process in Chapter 3. The problem is formulated as the definition of a global plan, δ, for a complex manufacturing system, which results from the composition of the single plans of the different manufacturing facilities making up the system. The target to be optimized is a bi-criteria cost function, $F(\delta) = [C(\delta), U(\delta)]$, which takes into account the production cost, $C(\delta)$, and a measure of the utilization of the facility, $U(\delta)$. The agents are associated with manufacturing entities at a highly aggregated level, that is, a whole plant or a production unit of a plant consisting of subcomponents, such as workshops, machines, products, : operations, orders, etc., managed within the plant or unit.

Each manufacturing facility or, equivalently, each agent, has a set of orders to produce that has been assigned to it in some separate way — for example, through the commercial channel that each facility holds with its "loyalized" customers or on the basis of some geographical location criteria. The key point is that each agent corresponds to an entity responsible for the P&S of its own set of orders. No matter how it is determined, the ith agent has a local plan, δ_i, and thus the global plan is given by

$$\delta = \bigcup_i \delta_i \, .$$

Clearly, having distributed *a priori* the set O_i of orders among the various agents, the whole performance achieved by δ° as the composition of the locally optimal (or close to optimal) plans $\delta_i^*(O_i)$ for the agents, namely

$$\delta^\circ = \bigcup_i \delta^*_i(O_i),$$

may lead to an inefficient (i.e., Pareto optimal) global solution.

Denoting a global efficient (Pareto optimal) plan with δ^*, it may result that $C(\delta^*) \le C(\delta)$ and $U(\delta^*) \le U(\delta)$, where at least one of the two inequalities is strictly satisfied. However, this does not mean that the P&S procedures used by the single agents are not effective, because there could be no chance of obtaining the global plan δ^* starting from the current partition of orders in the sets O_i. In fact, if a different order distribution, say O_i', among the agents were used, the plan

$$\delta^* = \bigcup_i \delta^*_i(O_i')$$

could ultimately be obtained. The MAS proposed by Frankovic and Dang [18] aims at discovering one such efficient partition by making the single agents cooperate.

Several possibilities or reasons can induce an agent to seek cooperation. One agent may send a request to others in order to receive help in completing its orders, that is, it tries to exchange, or subcontract, part of its plan to another agent. Conversely, an agent can propose itself to provide help in completing other plans because it has some un- or underutilized processing resource. The purpose of exchanging plans is thus to redistribute orders among manufacturing facilities following a sort of local search perturbation technique of the current initial solution. To do this, the agents must also exchange information about their plans. In this case as well, a number of alternatives about the information to be exchanged and when this can be done are available. For example, a trivial, poorly efficient possibility is that of all the agents always exchanging their complete plans with each other; another is that of allowing an agent to communicate part of its plan only whenever it is able to receive and process a new order.

A wiser approach entails taking into account possible future requests that can arrive from the other agents when an agent is building its plan (for example, by exploiting historic information about past requests). The exchange of part of a plan from one agent to another is ruled by a negotiation protocol similar to the contract net paradigm: a request corresponds to a part of a plan to be exchanged and to a payment offered for the service. This reduces the production cost of the agent accepting the exchange, thus representing an incentive to cooperate. Agents can also be clustered to facilitate the communication and the exchange of parts of their plans with other groups of agents. In this case, the resulting

MAS architecture is a hierarchy of homogeneous agents, and the global optimization is the consequence of the cooperation between single agents and clusters of them.

A first possible remark about this model is that it does not specify how the single agent can define its own plan, and this is an admittedly complex task. A second remark regards the fact that the single plans (and thus the global one) are basically defined off-line or, to be more precise, depend on the kind of P&S procedure, classic or agent based, off-line or on-line, used at the single agent level.

An example of a functional decomposition and cooperation approach for dynamic scheduling is given by Kouiss et al. [19]. The authors introduce agents as local real-time schedulers for a specific work center; their function is to select the most appropriate dispatching rule from among a set of predefined ones to assign the incoming jobs to that center's resources. The agents observe the system state and the possible occurrence of events (e.g., the availability of a new machine or the arrival and completion of jobs) and modify the active rule in order to keep the center performance at an acceptable quality level.

The MA system is composed of a two-layered hierarchy, as depicted in Figure 6.5. At the higher level, a supervisory agent monitors the global behavior of the MS and, in particular, can detect the occurrence of critical

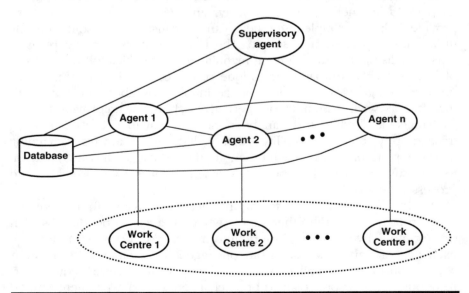

Figure 6.5 Multiagent Architecture (Redrawn from Kouiss, K. et al., *J. Intelligent Manuf.*, 8, 41, 1997)

situations, such as an excessively extensive deviation of the global performance from the manufacturing objectives. A number of local agents associated with a single work center operate at the lower level. These agents use their own knowledge base, implemented with a set of production rules, to detect local critical conditions and to react to them by selecting the most appropriate dispatching rule. The supervisory agent can influence the behavior of the local agents, thereby making them revise their current dispatching rule to improve global performance. Thus, this approach is an example of exploiting agents to decentralize decisions in order to gain flexibility. Different kinds of production orders can be scheduled by local decision-makers with different dispatching policies; this can be continuously adapted to new situations that emerge. Clearly, this is a functional decomposition approach in which the scheduling function is locally distributed.

Sikora and Shaw [20] proposed an MAS framework for the integrated management of all MS components and operations. Using the real-world application of the manufacture of printed circuit boards, they devised an MAS for scheduling a flow line with sequence-dependent set-up times. Belonging to the class of NP-hard problems, this kind of problem is usually approached by means of classic heuristics whereby it is decomposed into a pair of subproblems corresponding to (1) the definition of the lots to be produced by the line (the lot-sizing subproblem); and (2) the determination of the order to be followed for the production of the lots (the sequencing sub-problem).

The authors acknowledge that the lot-sizing and sequencing phases influence each other because of their contrasting objectives (i.e., in order to reduce the set-up costs, the lot-sizing tends to produce a few big lots, whereas in order to improve the completion times, the sequencing is favored by a large number of small lots), even if the classic approaches usually consider the two subproblems as independent. Thus, they define an agent-based framework in which the scheduling function is decomposed into two integrated decision modules corresponding to a lot-sizing agent and a sequencing agent. The schedule for a flow line derives from the cooperation between the two agents. The lot-sizing agent computes tentative lots on the basis of information about the current bottleneck machine and the makespan coming from the sequencing agent; on the other hand, the sequencing agent computes the makespan (by means of a local search heuristic) and determines the bottleneck machine on the basis of the size of the lots received from the lot-sizing agent. Actually, the iterative collaboration process is driven by the lot-sizing agent; it generates a sequence of different lot configurations and evaluates their performance, depending on the makespan provided by the scheduling agent and using a utility function that aggregates the two objectives to be minimized (the makespan and the set-up and holding costs).

Sikora and Shaw show how to extend this collaborative scheduling procedure from the level of a single production line to one in which more production lines devoted to the production of different finished product components are present. They thus introduce two high-level agents, corresponding to the manufacturing and assembly stages, that can dynamically modify the priority of each product type in order to smooth variations that may occur due to unexpected events in the production of the different lines.

Hierarchical Models

A successful application of an MAS based on a hierarchical decomposition of the planning function is represented by the ProPlanT (production planning technology) system proposed by Marik et al. [21]. This planning system, based on a functional decomposition to identify the agents' roles, was designed for project-oriented rather than manufacturing-oriented production systems, i.e., for systems in which the design, customization, and assembly aspects of production are prevalent.

A fully functional prototype of ProPlanT was developed to support the production planning of TESLA-TV, a Czech company producing radio and TV broadcasting stations. The system is composed of a community of agents assumed to reproduce the organization of the company in departments or teams within the MAS. The functions that have been associated with the agents are quite similar to those characterizing the three hierarchical layers responsible for deciding and managing the production operations in an MS: the planning, scheduling, and execution control layers. Each functional layer is populated by a number of peer agents that can take charge of the production of a single order. Thus, the MAS system is designed to be project oriented when a project corresponds to the production of a single order. The presence of multiple agents on each level leads to a horizontal functional decomposition, as more agents at the same level are able to perform the same functions.

The ProPlanT system basically includes four classes of agents, as reported in Figure 6.6 (redrawn from Marik et al. [21]). The production planning agents (PPAs) operate at the highest layer because they are responsible for defining an order's plan, taking into account the aggregated availability of the production facilities. PPAs define the suitable set of partially ordered tasks that satisfy the order requirements. At the intermediate layer are the production management agents (PMAs), which manage the entire production process for the orders assigned by the PPAs; in particular, the PMAs must identify how the tasks making up a project can be effectively subdivided and assigned to the agents at the lowest layer,

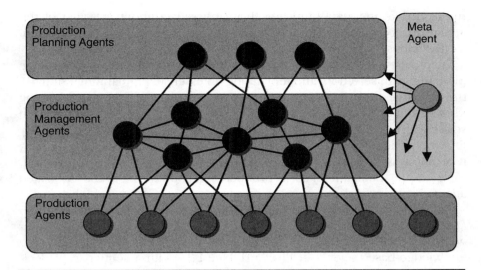

Figure 6.6 Three-Layered Architecture of the ProPlanT System (Redrawn from Marik, V. et al., *Int. J. Appl. Artif. Intelligence*, 14, 727, 2000)

i.e., the production agents (PAs), or subcontracted to other PMAs. Finally, the PAs are associated with the single production departments or operational units in the company, as well as with an encapsulated application. In particular, PAs can be scheduling agents in charge of scheduling the assigned tasks on the set of parallel machines of competence or database agents acting as front ends to the company database systems. Another agent, the so-called meta-agent (MA), is transversally introduced to monitor the behavior of the agent community, as well as to keep the knowledge bases used by the various agents updated. A special kind of monitoring agent, the replanning agent (RPA), is responsible for detecting failures and invokes the start of a replanning process (e.g., by reallocating the tasks that can no longer be processed by a facility).

The plan for a project (i.e., an order) derives from its successive definition and decomposition into tasks performed by the agents at the first and second layers of the hierarchy; these tasks are then subcontracted to other agents until all the production operations have been assigned to the PA. Apart from the PAs, the agents' activity is knowledge driven because PPAs, PMAs, and MAs operate as expert systems for their respective functions and roles. In particular, the PMAs use the so-called *tri-base acquaintance model*, that is, they include three distinct knowledge bases to store information and rules: the cooperator base; task base; and state base. The information stored in the knowledge bases can be static, like that included in the cooperator base, which allows identifying the other cooperating agents and establishing communication with them, or dynamic, like the state base, which includes data about the state of the agent's project or task. The task base contains the rules used by an agent to decompose and subcontract a task.

A project is processed by ProPlanT through four stages: preplanning; planning; plan selection; and replanning. In essence, preplanning estimates the resource and time requirements for the project, whereas planning identifies and continuously reviews possible alternative plans for it. Plan selection then determines which plan among the alternatives can actually fit into a schedule so that the corresponding production can start. Finally, replanning is invoked in case of changes occurring after plan selection, for example, the failure of some production resource (i.e., of the associated agent) or the arrival of a new project to be processed with high priority.

Ultimately, as can be seen, adopting a hierarchical decomposition approach means populating each organizational unit and functional layer with different classes of agents specialized for a specific activity. The knowledge-based agents in ProPlanT face P&S with a sequence of decisions throughout a three-layered hierarchy. PPAs and PMAs progressively reduce the complexity of a global monolithic P&S by distributing the workload between PMAs and PAs, respectively. This behavior reflects an optimization strategy that is certainly more flexible than the classic hierarchical layered approach used in ERP systems. For example, the possible subcontracting performed by PMAs could be considered a local optimization policy similar to the one followed in the coordination-based model analyzed in the section concerning agent-based applications in manufacturing planning and scheduling in Chapter 3. However, the main drawback of fully functional decomposition is the risk of overly embedding the decision-making logic in the agents instead of allowing it to emanate from the global behavior of the MAS. For example, it is evident that production-scheduling agents can operate, even if on a reduced subset of the whole problem, exactly the same as a scheduling procedure in a classic ERP system.

SUCCESSFUL MAS APPLICATIONS IN MANUFACTURING SCHEDULING AND CONTROL

As mentioned in Chapter 3, from a control standpoint, current successful MAS applications are generally referred to as holonic systems. Generally speaking, standards are needed at this level to control the shop floor processes properly; as a consequence, a working indicator of whether an MAS/holonic application is successful or not is to observe if it can potentially constitute a standard.

From this standpoint, as already stated in Chapter 3, the IEC 61499 standard [22, 23] seems to be one step up on the others, and applications based on this standard are flourishing. An example is the DCOS-1 architecture [24] designed to meet the basic requirements of metamorphic, i.e., dynamically adaptable and reconfigurable, control. One of the main advantages of the IEC 61499 standard is the combination of hierarchical and decentralized control in a unifying framework strongly based on software reusability in which software components encapsulate S&C knowledge directly on the devices. This is the first step toward allowing an agile configuration of the plant as far as S&C requirements are concerned. However, the IEC 61499 standard does not address the higher level requirements of cooperation, communication, negotiation, and, in general, high-level decision-making, despite preliminary attempts to the contrary. For example, Marik et al. recently proposed a general architecture combining function blocks with agents [25], in which a software agent and function block control application are encapsulated into a single structure called a holonic agent.

On the other hand, it can be said that the IEC 61499 standard effectively takes into account the many research efforts that have been made in the agent-/holon-based S&C field, such as hierarchical/decentralized control and encapsulation of S&C capability directly on the device. In fact, the combination of hierarchical and decentralized control is also emphasized by Fisher [26] as an effective solution for the design of FMS according to the holonic manufacturing paradigm. Fisher proposes the realization of holonic manufacturing systems (HMSs) by introducing agents at the different levels of the well-established hierarchical architecture for an FMS (depicted in Figure 6.7) composed by four layers: (1) *production planning and control* (PPC); (2) *shop floor control* (SFC); (3) *flexible cell control* (FCC); and (4) *autonomous systems* (AS). In particular, the benefits of MASs in S&C can be gradually introduced in this architecture starting from the lower level. In fact, PPC and SFC can conserve, respectively, the classic roles of defining highly aggregate production plans and of defining an off-line schedule and tracking the production process according to the information coming from the lower levels. Moreover, PPC and SFC can also be implemented as a set of agents/holons equipped with an appropriate P&S capability.

The AS layer corresponds to an MAS made up of agents with a DBI internal architecture (the InteRRaP agent architecture [27]) and, in particular, capable of local problem solving and social interaction for coordination. AS agents are associated with manufacturing resources and devices and are physically grouped into flexible cells representing production units of resources and devices working together; in addition, these agents can be logically grouped according to their function (mobile manipulation, transportation, machining). The description of the HMS in Fisher [26] focuses on the AS layers; AS agents receive from the SFC layer the

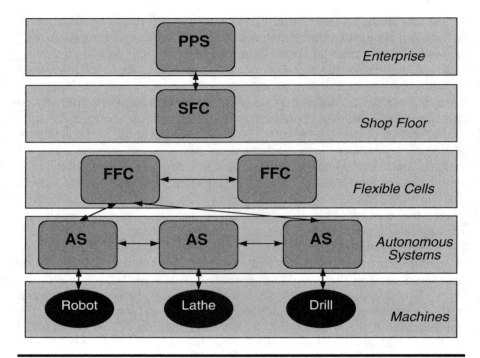

Figure 6.7 Planning and Controlling Hierarchy in FMS Where InteRRaP Agents Are Introduced to Implement HMS (Redrawn from Fisher, K., *Robotics Autonomous Syst.,* **27, 3, 1999)**

communication of the next tasks to be performed following the predefined schedule, but the final decision about the operation sequence is made at the MAS level.

Because each task actually needs more than one resource to be executed, AS agents coordinate their decisions to form a so-called *complete team* for every task. In the coordination adopted, the AS agents basically first try to assume full responsibility for the execution of a task, i.e., become the *leader* for it, then ask the other AS agents to join the necessary team. The agents thus become the channel through which the control commands are sent to the lowest level physical layer of the FMS hierarchy. Trials have confirmed the effectiveness of the approach and, in particular, the author points out the importance of combining the classic hierarchical division of the decisional layers with the distribution of detailed control to local autonomous decision makers, the AS agents.

Rabelo and Camarinha–Matos [13] highlight the need to develop encapsulating layers around the physical controllers, thereby hiding their specificity and diversity. In their work, the authors emphasize the interaction between scheduling and control and the need for strict integration. More specifically, they propose an MAS architecture oriented toward dynamic scheduling, indicating with this term the ability to react promptly to on-line disturbing events in a harmonic and smooth way. The architecture of the MAS proposed by these authors has already been discussed in the section on auction-based models and depicted in Figure 6.5; here, the same architecture is detailed in order to call attention to the role of the agents involved in the S&C decisions.

High-level scheduling corresponds to the decisions made by a *scheduling supervisor agent* (SSA), which assigns each single job to a cluster of resources, depending on resource availability and taking into account production parameters such as flow time, lateness, work in process, and utilization. The agents called *Consortia* (C) operate at an immediately lower level; these agents are dynamically created to execute a specific job and are associated with the cluster of resources assigned to the job. At the lower level are the *enterprise activity agents* (EAA), which are associated with the local controller of the single resources and whose responsibility is to execute the tasks and continuously monitor for local failures. A consortium thus represents a logical composition of a group of EAA formed according to several criteria (for example, the topological distance among the resources) and can be thought of as a virtual cell. Consortia live only the time needed to execute the jobs, are responsible for the completion of the jobs, and have the decision capacity to reschedule locally, for example, in case of a machine breakdown. Detailed device control on the other hand, is performed by the EAA.

The *local spreading center* agents (LSC) reside, like the consortia, at the intermediate level; these agents do not have decision capability, but operate as facilitators, thus allowing efficient communication between the SSA and the EAA. The LSC are associated with clusters of functionally homogeneous resources, and their presence obviates the need to broadcast the announcement of tasks to the whole EAA community, limiting the communication only to clusters providing the required kind of service. The SSA determines the schedule, first announcing the jobs to the EAA via the LSC, and then assigning the execution of the job's operations to compatible EAAs that declare themselves available, selecting them on the basis of scheduling objectives, and forming in this way the consortia. Scheduling involves agents represented in Figure 6.3 as gray shaded boxes; agents implicated in on-line control activity are denoted by white boxes in the same figure. Control is mainly in charge of the consortia and the EAA, which can make only local decisions to react to possible faults, eventually attempting some local rescheduling before asking the SSA to intervene whenever a deeper rescheduling action is indispensable.

The physical device encapsulation is implemented at the EAA level. These agents are composed of two types of interacting processes: manager

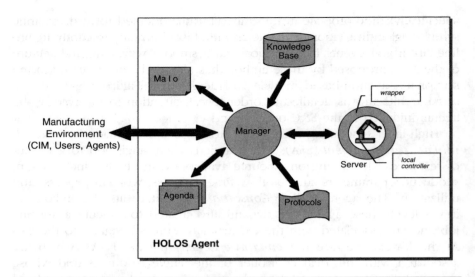

Figure 6.8   HOLOS Agent Architecture (Redrawn from Rabelo, R.J. et al., *Robotics Autonomous Syst.*, 27, 15, 1999)

and server; the former controls task execution and monitors possible deviations from the nominal local plan, whereas the latter corresponds to the physical controller and must report the device state in real time. Thus, the EAA inner architecture basically follows the client–server scheme. In addition, whenever an EAA is associated with a set of resources whose functions are very closely related (e.g., a fixed tool exclusively used by a machine), a single manager can also interact with multiple servers.

This model for agents that directly control devices has also been used by Rabelo et al. [12] in the more recent HOLOS system, in which agents are integrated with the enterprise information system. In HOLOS the agents are provided with user interfaces and can communicate with the production resources and the CIM information system by means of standard protocols, MAP/MMS, and STEP/DAI (Figure 6.8). Integration with shop floor resources, such as robots or NC machines, is achieved by introducing an encapsulation layer (wrapping) around the physical resource controller, which is able to communicate with the logical manager of the EAA.

The requirement for a so-called software wrapper to encapsulate legacy manufacturing device controllers is recognized by Barata et al. [28]. Three platforms for supporting the development of multiagent applications have been analysed: JatLite [29]; FIPA-OS [30]; and Jade [31]. The authors specifically compare the platforms in relation to several aspects, among which are the use of standards, support and documentation available, and supports for management of the agent community. Barata et al. conclude that FIPA-OS and Jade are quite similar, but that the second is preferable, given the documentation available and the tested rapidity in obtaining the first applicative results from implementation. Jade, in fact, satisfies all the desired requirements for a development system because it is FIPA compliant, well-supported even with a mailing list, and includes tools to manage and supervise community activities, e.g., *directory finder* agents that provide a yellow pages service and a *remote monitoring agent* that controls the life cycle of the agents.

The availability of appropriate technologies and of a reliable agent platform provides developers with the basic building blocks for the design and implementation of agent-based S&C systems. One pivotal aspect that remains, however, involves interaction with physical devices: different devices may require specific communication mechanisms because they are usually provided with proprietary protocols and different functionalities and specifications. Thus, an interface module is needed that converts the specific interaction protocols and fits them to physical devices so that their logical controllers, i.e., the agents, can monitor and act on them. Barata et al. [28] point out the importance of using standards in implementing the interaction layer between the physical and logical part of a control system, and they propose a valid solution for the development of object libraries based on CORBA or DCOM communication architectures. Basically, the objects in the libraries should expose methods for reading and writing variables, for manipulating programs (such as download, upload, start and stop), and for being notified on the occurrence of events.

Application of these concepts are illustrated in the control of the Novaflex flexible assembly cell installed in the Uninova Institute facility, the Institute for the Development of New Technologies in Lisbon, Portugal. The cell is composed of four subsystems: two assembly cells, an automated warehouse, and a transportation system. Both the assembly cells are based on two robots of different brands and include a tool exchange mechanism and a fixing device installed on the conveyor in front of the robot. Therefore, the flexible cell integrates many devices equipped with heterogeneous controllers and represents a suitable case study in which a number of possible interaction difficulties due to nonstandard communications are a benchmark for the effectiveness of an agent-based control system. The multiagent approach followed by the authors is characterized by two features: the *agentification* of the manufacturing components and the establishment of an agent community.

The latter corresponds to the definition of two types of organizational structures similar to Rabelo et al. [12], i.e., *clusters* and *communities*. A manufacturing cluster is a long-lasting organization associated with a group of agents connected to physical components, whose purpose is to organize and share the common aspects linking the components, such as the ontology; services and skills; communication protocols; and mechanisms, that allow it to be informed about job opportunities. A consortium is a dynamically formed temporary group of agents created to induce inter-agent collaboration in order to achieve a specific objective. The process of agentification entails connecting the physical controller to the associated agents. The integration problems arise especially when legacy systems that centralize the control of many physical devices are present. In these cases, a software wrapper should be developed to encapsulate the details of each device, as schematized in Figure 6.9 (redrawn from Barata et al. [28]). The agent perceives the presence of an abstract device through the wrapper and can remotely interact with the wrapper, using the methods exposed by a local proxy interface of the MAS and ultimately communicating with the physical controller.

Focusing now on the possible architectures of MASs applied to S&C in manufacturing, an alternative, which has often been considered a counterpart of the semihierarchical architecture proposed by Rabelo and Camarihna–Matos [13], is the market-based architecture by Lin and Solber [32], who propose a model based on an autonomous agent architecture in which S&C capabilities are decentralized to the single entities affected in the manufacturing shop floor. Two main types of agents, i.e., job agents and resource agents, are introduced following a physical decomposition, and the S&C decisions are made according to the contract net-based negotiation protocol reported in Figure 6.10.

The job agents are transient and are created whenever a new job to be manufactured enters the shop floor. A job agent holds the knowledge about the processes needed for the execution of the physical job, retains several alternative plans that can be followed, and knows the set of weighted objectives that should be reached. The job agent includes:

- A *data file* to store the information about the job and a fictitious fixed budget that takes into account its objective priorities.
- An *intelligent file* containing the procedures used during the negotiation, together with the state of the current bid offered to the resource.
- An *inferential engine* that controls the global agent behavior.

A resource agent is permanently associated with a simple physical device or an aggregation of them. Even the resource agents include a data file to store information about the queue length, available capacity, state of the processing and the bids received, and an inferential engine ruling the agent behavior. The S&C process evolves on-line and is based on a market social model. Before its current operation is completed, the job agent announces the task for the next operation to the resources, specifying its characteristics and a bid of virtual money. The resource

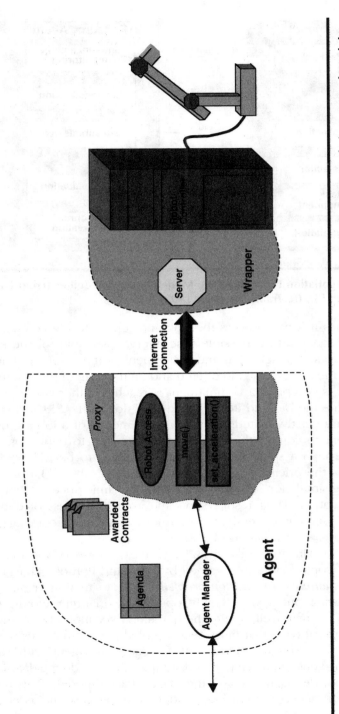

Figure 6.9 Connection of an Agent to a Physical Device through a Wrapper (Redrawn from Barata, J. et al., in *Proc. 3rd Workshop Eur. Sci. Ind. Collaboration*, Enschede, The Netherlands, 2001)

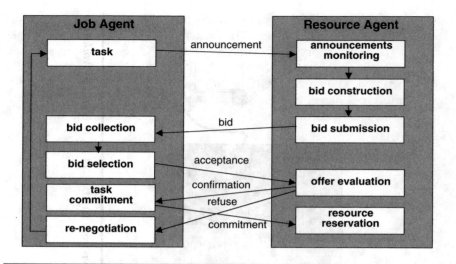

Figure 6.10 Negotiation Protocol of the Model by Lin and Solber (From Lin, G.Y. and Solberg, J.J., *IIE Trans.*, 24, 169, 1992)

agent determines if it can process the task and replies to the compatible requesting job agent with a possible time interval available for the execution of the tasks. The selection and assignment of the resource time is determined on the basis of the bid offered and on the state of the resource. If the proposal is accepted, a commitment is reached; otherwise, the job agent can update the bid and iterate the negotiation cycle. S&C activities in this model are strictly related; the task requirements for a job are used by the job agent to control the job's manufacture and to define on-line its schedule through a stepwise selection of the plan to be followed. A critical aspect is the pricing policy adopted. Lin and Solberg [33] present a manufacturing simulation system based on the dynamic price mechanism for agent negotiation. Their results show the appropriateness of dependence on priority and, in general, the robustness of the architecture against resource failures or variations in objectives.

The model by Lin and Solberg [32] was the premise of the on-line scheduling and control system described by Brun and Portioli [34] to face the so-called *assembly coordination problem*. Such a situation entails not only manufacturing but also assembly operations; in other words, the production of a job may well require the (parallel) execution of operations on different physical parts that must be assembled at certain stages of the process. Brun and Portioli's model [34] associates an agent with each resource and an agent with each part and compels the scheduling decisions to emerge from a negotiation (as in the market-based model). This time, however, the part agents are not independent and, because the assembly operations correspond to the appropriate matching of them, part agents need a coordination mechanism. This is achieved by associating a black-board (a shared memory) with each assembly operation, where updated information about the part processing state is shared among all the part agents involved. The part agents can thus better estimate the time at which the assembly operation can take place and can tune the bid offered to the resource agents accordingly (i.e., the time slot required and the virtual money tendered) — not only in order to meet their due dates but also to synchronize dynamically with the partners in the assembly.

The different multiagent architectures proposed by Rabelo and Cama-rinha–Matos (semihierarchical) on the one hand and by Lin and Solberg (market based) on the other are two main representative approaches to the decentralization of S&C in manufacturing. The experimental comparison of these architectures performed by Cavalieri et al. [35] and the conclusions ultimately drawn by the authors thus provide some intriguing food for thought. The behavior of the two models was simulated using a fictitious manufacturing system made up of different types of machines that can produce several finished products but require different numbers and kinds of operations. The authors simulated several product mixes, considering sequences of jobs with a fixed or variable (i.e., not predefined) order of operations, and sequences of jobs with a combination of the two cases. The experimental campaign analyzed three situations: a stable production load (i.e., the jobs are released as planned at the beginning of each shift and, even if the processing time can stochastically vary, no plan disturbance occurs); a production load with arrivals of urgent jobs; and a production load with machine breakdowns. The performance parameters considered were the mean flow time, the delay time, and the number of late jobs.

The results highlighted a behavior that was, to a certain extent, contrary to intuition: the market-based architecture seemed particularly oriented toward the optimization of the mean flow time and presented a better performance in the case of stable production load with a homogeneous type of mix (i.e., fixed or variable but not combined). On the other hand, the semihierarchical model in general yielded results with a lower disper-sion around the averages computed and behaved better with delay-related objectives and, particularly, in cases in which unexpected events (urgent jobs or machine breakdowns) could occur. The market-based architecture was able to respond promptly to unexpected events because it always implements a local, job-centered optimization, which limits the propaga-tion of perturbations. Agents associated directly with the jobs handled the rapid flow of job operations in the system and effectively react to stochastic variations in the processing times, thus explaining the good result in the mean flow time. However, in the presence of disturbances, the agents penalized in the negotiation often completed the jobs with relevant delays.

The global optimization capability of the market-based model derives from the pricing scheme used. The information about the job priority is, in fact, implicitly taken into account by assigning to the jobs a budget of virtual money proportional to their priority, thereby encouraging them in the negotiations. The performance of the semihierarchical model improved for nonstable production loads because, in these cases, the role of rescheduling played at the global level by the supervisor produced the best results: whenever the problems caused by a disturbance could not be locally controlled by the EAA, the SS had the chance to revise the future scheduling decisions, thus smoothing out the effects of variations. Even if the market-based model outperformed the semihierarchical one in the case of the mean flow time, the latter did not excessively delay the jobs and generally provided more stable performances.

As concluding remarks to this section, it is possible to underline the accurate observations by McFarlane and Bussmann [36]. S&C systems offer clear advantages in adopting a decentralized decision approach, particularly a multiagent or holonic one. A hierarchical distribution of the decision responsibilities among agents seems at present to be the more reliable alternative to exploit such advantages; the decentralization at the lower level of the hierarchy allows unexpected schedule variations to be faced locally, but the supervision of higher decision levels guarantees the ability to redefine the schedules with a global system view. However, the price to be paid for this good and stable performance even under unstable working conditions is a poorer responsiveness and reconfigurability. Although pure autonomous agent architectures for S&C can clearly respond very rapidly to changes, something more sophisticated is still needed in terms of cooperation mechanisms to overcome the limits deriving from narrow local decision-making. On the other hand, because they are easily reconfigurable and scalable, autonomous agent solutions enjoy high decentralization.

PRACTICAL WORKING INDUSTRIAL APPLICATIONS

Despite the extraordinary promise of agent-based approaches to agile manufacturing systems, only a few practical working industrial applications have been developed. Even fewer can actually completely fit the real manufacturing world. According to a recent review by Shen [37], IBM was one of the first companies to implement a testbed industrial application, adopting the logistics management system (LMS) in commercial production [38] that used functional agents, one for each production constraint, and a judge agent to combine the votes of four different critics.

Other testbed applications that can be more aptly related to the manufacturing world are found in the reviews of several outstanding authors, all of them witness to the disappointing gap between theoretical models and practical applications and all intent to forge a solution to reduce this gap. For example, Parunak has always been mindful of practical industrial applications, and his Web site [39] should be viewed as an excellent source for a critical description of the current state of the art of agent-based manufacturing.

Jennings and Bussmann [40] recently described a manufacturing line control testbed application, focusing on the problem of finding an alternative to centralized and preplanned scheduling approaches that can rarely be respected. To identify a flexible agile solution, the authors developed the concept of a modular manufacturing system, stating that the entire manufacturing system should be composed of standard modules. In this approach, a first-price, sealed auction is used to allocate workpieces to machines, and a specific agent is associated with each workpiece, each machine puts/gets parts on/from a conveyor, and each transport switch moves parts from one conveyor to another. This system was initially simulated by DaimlerChrysler in order to evaluate the robustness of the system when it is subject to disturbance and to evaluate its performance: the system achieved a rating of about 99.7% against the theoretical optimum. In a second phase, the control system was installed as a bypass line to an existing large-series manufacturing line for cylinder heads in a plant located in Stuttgart–Untertürkheim, Germany. The system has been running for routine operations for more than 2 years, thus confirming its robustness.

Even more recently, new paradigms based on Jade [31] technology have been devised and applied to the world of manufacturing. Apart from the MAST simulation tool by Vrba [41], which currently seems to hold more potential when run on standard PLC-based automation controllers (Rockwell Automation ControlLogix platform) in parallel with the low-level, real-time control code (ladder logic), a Jade-based testbed application is described by Becvár et al. [42] (see also Pchouek et al. [43]). Specifically, their ExPlanTech/ExtraPLANT solution was applied in the first ExPlanTech version to Liaz Pattern Shop (Czech Republic), which produces forms and patterns for the European automobile industry, and to Hatzapoulos (Greece), a producer of packaging technology. In addition, it is expected to be used to plan production at the new engine assembly plant of SkodaAUTO, the VW-owned car manufacturer.

CONCLUSIONS

In this chapter, a wealth of ideas and experiences on how to "agentify" a manufacturing enterprise in order to make it agile has been presented. The literature considered is just a sample of the many research efforts in this field. The authors would take this chance to apologize for neglecting many important projects in this chapter and, in general, throughout this book.

The applications were presented, following in part the structure of Chapter 3, thereby aiming to distinguish between P&S and S&C applications and always taking into account the layered architecture of current manufacturing information systems. No specific framework has been used to compare the different architectures and projects because their evaluation goes beyond the aim of this book. Moreover, given the current "primordial era" of agent-based manufacturing, any attempt at comparison would have made for a hazardous exercise. In any case, the correct agent-based solution is still a matter subject to the specific needs of a manufacturing enterprise. However, readers interested in a practical classification of agent-based technology may refer to Parunak [44], who describes the technology according to its functions in the life cycle of the industrial system and according to the following key descriptions:

■ Agent mapping, in relation to the types of agents and their functions: for example, resource agent, work order agent, etc.
■ Agent modeling, in relation to modeling the knowledge of the agent and of other agents in the system: for example, the use of a community-wide blackboard to share the knowledge of the current working state of each agent
■ Agent structure, in relation to some salient characteristics of the agent as an algorithmic component: for example, whether it can modify its structure over time and whether it maintains a memory of previous states
■ Population, related to a numeric evaluation of the agents present in the system, possibly related to real testbed applications
■ Communication channels, related to the description of how two agents can communicate with each other
■ Communication protocols, mainly related to the semantics of the messages among agents: for example, whether they use a contract net-based protocol
■ Configuration, mainly related to the description of interaction among agents in the MAS systems: for example, whether they are fixed in design or they can be set up dynamically in run-time
■ Coordination, mainly related to the flow of information within the MAS: for example, coordination may result as the propagation of the constraints of the bidding process among the agents
■ Maturity, mainly related to the presence of testbed applications

As far as this last aspect is concerned, one key point that has arisen from the review in this chapter is that few real working, agent-based applications of agile manufacturing systems have been implemented throughout the world; moreover, those in place were put there by outstanding organizations and applied to important industrial realities. With this premise, it is admittedly difficult to convince PS-Bikes' decision-makers that the agent is the right solution for them. By the same token, it should be borne in mind that small and medium enterprises (SMEs) constitute the principal European industrial structure, in numbers and in employees [45], and that agility is a characteristic that has made SMEs a successful production model.

Paradoxically, however, SME agility is now jeopardized because of the overwhelming flood of information needed to accomplish modern production. To solve this impasse, a dual approach is required. First, holonic manufacturing should be introduced in current machinery standards, with the overriding goal to enhance vertical integration among the different layers of the manufacturing information system. Second, agent-based management procedures, applied to workflow management, for example, should be introduced as a natural extension of traditional desk-top office software applications. Here, the objective is to enhance horizontal integration with special reference to supply chain management and to allow the creation of networks of small–medium virtual enterprises, which can better react to changes in market demands.

In fact, authors of recent works [41–43, 46] seem to be well aware of the need for MAS solutions for intra-enterprise production and extra-enterprise activities that can be applied equally well to SMEs and to large manufacturers in order to achieve rapid due-date response; a high degree of manufacturing flexibility; minimization of stock resources; a balanced load of the manufacturing machinery; etc. In lieu of standardized approaches toward agentification and enhanced holonic manufacturing, for the moment the IEC 61499 standard seems to provide the best working alternative.

REFERENCES

1. Cho, H., Jung, M., and Kim, M., Enabling technologies of agile manufacturing and its related activities in Korea, *Computers Ind. Eng.*, 30, 323, 1996.
2. Gunasekaran, A., Agile manufacturing: a framework for research and development, *Int. J. Prod. Econ.*, 62, 87, 1999.
3. Gunasekaran, A., Agile manufacturing: enablers and an implementation framework, *Int. J. Prod. Res.*, 36, 1223, 1998.
4. Shen, W. and Norrie, H.N., An agent-based approach for dynamic manufacturing scheduling, in *Proc. Autonomous Agents'98 Workshop Agent-Based Manuf.*, Minneapolis/St.Paul, MN, 117, 1998.
5. Shen, W. and Barthès, J.P., An experimental multi-agent environment for engineering design, *Int. J. Cooperative Inf. Syst.*, 5, 131, 1996.
6. Gaines, B.R., Norrie, D.H., and Lapsley, A.Z., Mediator: an intelligent information system supporting the virtual manufacturing enterprise, in *Proc. 1995 IEEE Int. Conf. Syst., Man Cybernetics*, New York, 964, 1995.

7. Maturana, F., Balasubramanian, S., and Norrie, D.H., A multi-agent approach to integrated planning and scheduling for concurrent engineering, in *Proc. Int. Conf. Concurrent Eng.: Res. Applications*, Toronto, Ontario, 272, 1996.
8. Shen, W., Norrie, D.H., and Kremer, R., Developing intelligent manufacturing systems using collaborative agents, in *Proc. 2nd Int. Workshop Intelligent Manuf. Syst.*, Leuven (B), 157, 1999.
9. Parunak, H.V.D., Baker, A.D., and Clark, S.J., The AARIA agent architecture: from manufacturing requirements to agent-based system design, *Integrated Computer-Aided Eng.*, 8, 45, 2001.
10. Parunak, H.V.D., Sauter, J., and Clark, S.J., Toward the specification and design of industrial synthetic ecosystems, *Intelligent Agents IV: Agent Theories, Architectures, and Languages, Lecture Notes in Artificial Intelligence*, Singh, M.P., Rao, A.S., and Wooldrige, M., Eds., Springer, Berlin, 45, 1998.
11. Gu, P., Balasubramanian, S., and Norrie, D.H., Bidding-based process planning and scheduling in multi-agent system, *Computers Ind. Eng.*, 32, 477, 1997.
12. Rabelo, R.J., Camarinha-Matos, L.M., and Afsarmanesh, H., Multi-agent-based agile scheduling, *Robotics Autonomous Syst.*, 27, 15, 1999.
13. Rabelo, R.J. and Camarinha–Matos, L.M., Negotiation in multiagent based dynamic scheduling, *J. Robotics Computer Integrated Manuf.*, 11, 303, 1994.
14. Tuijnman, F. and Afsarmanesh, H., Management of shared date in federated cooperative PEER environment, *Int. J. Intelligent Cooperative Inf. Syst.*, 2, 451, 1993.
15. Sousa, P. and Ramos, C., A distributed architecture and negotiation protocol for scheduling in manufacturing systems, *Computers Ind.*, 38, 103, 1999.
16. Liu, J.S. and Sycara, K.P., Coordination of multiple agents for production management, *Ann. Operations Res.*, 75, 235, 1997.
17. Archimede, B. and Coudert, T., Reactive scheduling using a multi-agent model: the scep framework, *Eng. Applications Artif. Intelligence*, 14, 667, 2001.
18. Frankovic, B. and Dang, T.T., Cooperating agents for planning and scheduling, in *Proc. IFAC Workshop Manuf., Modelling, Manage., Control — MIN 2001*, Prague, 8, 2001.
19. Kouiss, K., Pierreval, H., and Mebarki, N., Using multi-agent architecture in FMS for dynamic scheduling, *J. Intelligent Manuf.*, 8, 41, 1997.
20. Sikora, R. and Shaw, M.J., Coordination mechanisms for multiagent manufacturing systems: application to integrated manufacturing scheduling, *IEEE Trans. Eng. Manage.*, 44, 175, 1997.
21. Marik, V., Pechoucek, M., Stepankova, O., and Lazansky, J., ProPlanT: Multiagent system for production planning, *Int. J. Appl. Artif. Intelligence*, 14, 727, 2000.
22. Christensen, J.H., HMS/FB architecture and its implementation, in *Agent Based Manufacturing: Advances in the Holonic Approach*, Deen, S.M., Ed., Springer Verlag, 53, 2003.
23. Lewis, R., *Modelling Distributed Control Systems Using IEC*61499. Applying Function Blocks to Distributed Systems*, IEEE, 2001.
24. Balasubramanian, S., Brennan, R.W., and Norrie, D.H., An architecture for metamorphic control of holonic manufacturing systems, *Computers Ind.*, 46, 13, 2001.
25. Marik, V., Pechoucek, M., Vrba, P., and Hrdonka, V., FIPA standards and holonic manufacturing, in *Agent Based Manufacturing: Advances in the Holonic Approach*, Deen, S.M., Ed., Springer Verlag, 89, 2003.

26. Fisher, K., Agent-based design of holonic manufacturing systems, *Robotics Autonomous Syst.*, 27, 3, 1999.
27. Fischer, K., Müller, J.P., and Pischel, M., A pragmatic BDI architecture, in *Intelligent Agents — Proc. 1995 Workshop on Agent Theories, Architectures, Languages (ATAL-95)*, Wooldridge, M., Müller, J.P., and Tambe, M., Eds, *Lecture Notes in AI*, Springer-Verlag, 1037, 203, 1996.
28. Barata, J. et al., Integrated and distributed manufacturing, a multi-agent perspective, in *Proc. 3rd Workshop Eur. Sci. Ind. Collaboration*, Enschede, The Netherlands, 2001.
29. JatLite, Web site http://java.stanford.edu.
30. FIPA-OS, Web site http://fipa-os.sourceforge.net.
31. Bellifemine, F., Caire, G., Poggi, A., and Rimassa, G., JADE. A white paper, *EXP — Search innovation*, 3, 6, 2003, available at http://exp.telecomitalialab.com
32. Lin, G.Y. and Solberg, J.J., Integrated shop floor control using autonomous agents, *IIE Trans.*, 24, 169, 1992.
33. Lin, G.Y. and Solberg, J.J., An agent-based flexible routing manufacturing control simulation system, in *Proc. 1994 Winter Simulation Conf.*, 970, 1994.
34. Brun, A. and Portioli, A., Agent-based shop-floor scheduling of multistage systems, *Computers Ind. Eng.*, 37, 457, 1999.
35. Cavalieri, S., Garetti, M., Macchi, M., and Taish, M., An experimental benchmarking of two multi-agent architectures for production scheduling and control, *Computers Ind.*, 43, 139, 2000.
36. McFarlane, C.D. and Bussmann, S., Development in holonic production planning and control, *Int. J. Prod. Control*, 11, 522, 2000.
37. Shen. W., Distributed manufacturing scheduling using intelligent agents, *IEEE Intelligent Syst.*, January/February, 88, 2002.
38. Fordyce, K. and Sullivan, G.G., Logistics management system (LMS): integrating decision technologies for dispatch scheduling in semiconductor manufacturing, in *Intelligent Scheduling*, Zweben, M. and Fox, M.S., Eds., Morgan Kaufmann, San Francisco, 473, 1994.
39. Parunak, H.V.D., home page, http://www.erim.org/~vparunak/
40. Jennings, N.R. and Bussmann, S., Agent-based control systems. Why are they suited to engineering complex systems?, *IEEE Control Syst. Mag.*, June, 61, 2003.
41. Vrba, P., "MAST: Manufacturing Agent Simulation Tool," exp — Vol. 3, No. 3, 2003, available at http://exp.telecomitalialab.com.
42. Bečvář, P., Charvát, P., Pěchouček, M., Posíšil, J., Řiha, A., and Vokřínek, J., "ExPlanTech/ExtraPLANT: Production Planning and Supply.Chain Management Multi-Agent Solution," exp — Vol. 3, No. 3, September 2003, pp. 106–115, available at http://exp.telecomitalialab.com.
43. Pěchouček, M., Říha, A., Vokřínek, J., Marík, V., and Praûma, V., ExPlanTech: Applying multi-agent systems in production planning, in *International Journal of Production Research*, Vol. 40, No. 15, 3681–3692, 2002.
44. Parunak, H.V.D., Practical and Industrial applications of agent-based systems, Environmental Research Institute of Michigan (ERIM), 1998, available at http://www.erim.org/~vparunak/.
45. Hvolby, H.H. and Trienekens, J.H., Special issue: stimulating manufacturing excellence in small and medium enterprises, *Computers Ind.*, 49, 1, 2002.

46. Montaldo, E., Sacile, R., and Boccalatte, A., Enhancing workflow management in the manufacturing information system of a small-medium enterprise: an agent-based approach, *Inf. Syst. Frontiers*, 5, 195, 2003.

7

FUTURE CHALLENGES

Having reached the conclusion of the book, we hope that the reader and the PS-Bikes staff will feel acquainted with the current state of the art of agent-based manufacturing and the related practical possibilities to enhance the agility of manufacturing companies in order to achieve peak performance. On the other hand, it should be evident that manufacturing "agentification" is still far from being "the" approach. For example, Chapter 3 and Chapter 6 highlighted the gap still separating conventional and agent-based planning, scheduling, and control techniques. It follows that the reader and PS-Bikes staff should welcome — as a conclusion to this book — greater insight into the current research trends in this field and into what should be expected in the next few years. This text is not meant to serve as an oracle of agent-based manufacturing's imminent future; however, this closing chapter will describe these issues, endeavoring to highlight what has yet to be done or otherwise refined in the field. Here, academic researchers should be able to find important indications about the most promising, but not fully explored, trends in agent-based research applied to the manufacturing world.

INTRODUCTION

Are agents and multiagent systems (MASs) a suitable solution for next-generation agile manufacturing systems? What direction should research on MASs take in the near future to enhance their appropriateness for manufacturing? These are the main questions that this final chapter seeks to answer. In this introduction, some recommendations arising from earlier chapters about the application of MASs in manufacturing are summarized; thereafter, particular attention will be devoted to the discussion of the future of MASs.

Let us first try to summarize some key aspects pointing to the suitability of an MAS approach in manufacturing. Manufacturing systems, which are very dynamic in establishing relationships with partners, suppliers, and customers as well as in defining new product designs, plans, and schedules, can no doubt find expedient management solutions in MAS applications. Therefore, the modern manufacturing requirements of flexibility and leanness on the one hand, and reactiveness and agility on the other, are stimuli for MAS applications. Consider, for example, two possible extremes: a manufacturing system (MS) made up of a single plant, characterized by quite stable make-to-stock production of a few relatively chip mass goods and a small consolidated number of suppliers. At the other extreme, consider an MS corresponding to a network enterprise that manufactures high-quality customized products on a make-to-order (assembly-to-order or design-to-order) basis as the result of the cooperation of a set of specialized industries with one or more plants each.

Decisions in the second context, whatever their level of detail, are clearly more complex because of the larger number of inputs to be considered and outputs to be defined; the number of interconnected entities whose activities must be coordinated in order to best reach the MS objectives; and the fact that plans are generally more crucial in the case of high-quality customized and expensive products, thereby making wrong decisions potentially very detrimental. The greater the complexity of a system to be managed is, the more inappropriate a monolithic centralized solution becomes, especially if it is expected to handle all aspects simultaneously, from supply management to shop floor device control. Heterarchical and semiheterarchical management solutions have been recognized as absolutely necessary to mitigate such complexity. MASs can be tailored to introduce the desired level of decentralization in MS.

Modern MSs must be agile; that is, they must be able to adapt to the dynamic changes of the world in which they operate. Due to their autonomy and proactiveness, MS components should be readily updatable and reconfigurable. Agents and/or holons allow such a reconfigurability. Dynamicity often requires on-line decision-making ability; agent-based decision-making, emerging from negotiation/collaboration protocols, is naturally exploitable in on-line scenarios. Dynamicity in modern manufacturing also relies on the availability of continuous flows of fresh information. Again, software agents can provide the solution because they introduce a lean communication channel among the entities in an MS, thus enabling interoperability. The uptake of agent technology can be gradual and requires relatively contained investments in hardware or software, so even the MSs of small–medium enterprises (SMEs) can move toward agility without remaining subject to the pitfalls stemming from an abrupt change in consolidated practices.

As pointed out in Chapter 3, in the complex dynamic scenarios of modern manufacturing, fundamental activities such as planning, scheduling, and control can be appropriately handled by MAS approaches. On the other hand, if an MS with simple relationships with suppliers and customers and stable production is satisfied by its current approach to planning, scheduling, and control, it is not necessary to abandon such tested terrain by seeking the same level of satisfaction through an MAS approach. MAS approaches do not promise the moon, and in general the opportunity of abandoning a management solution already in place should be carefully evaluated. However, consider a scenario in which the manufacturing processes and their management are not so stable and something will soon change within and around the MS that demands a more flexible, agile, robust approach to increasingly uncertain scenarios (e.g., the introduction of new information tools to access e-commerce opportunities; modified relations with suppliers and customers; or a different company strategy). Then, the benefits of the gradual introduction of an MAS into such a manufacturing reality must be seriously considered.

At this point, we are entitled to ask about the outlook for research on the applications of agent technology in manufacturing. Agents can be considered and studied from two standpoints: as autonomous software components or as autonomous entities of a complex software organism, i.e., the agent society. Autonomous agents, as business components, could become responsible for even more complex tasks, relieving human operators of routine or noncritical (bureaucratic) tasks (e.g., workflow component and manager agents). Organizations of autonomous agents could become a standard for establishing systems' interoperability. Agents could autonomously exchange information on behalf of the company to which they belong, maintaining contacts among partners that are always alive and updated. For instance, e-procurement activities, like the search for commercial suppliers or partners, could be highly streamlined by a worldwide network of marketing agents with a common communication TCP/IP-based protocol and market-place facilitators. Business component agents, e.g., data mining agents, can perform autonomous data analysis, navigating in the Web or mining a company's archives, in order to provide executives with analytical information needed for strategic decisions proactively.

Because agents, from the simplest to the most complex, can be organized into societies, they could be viewed as the building blocks of new organization models for manufacturing industries. In fact, an industry may be viewed as an organism that lives in a given world with rules, e.g., the market, responds to the world's stimuli, and acts in order to survive and grow. The industry's components, such as resources and functions, should be internally structured to reach that goal. However, given the constant changes in the world market, there is no definitive answer as to what the best structure is. Therefore, agents associated with a company's physical and functional components could be allowed to seek out their own self-organization freely and to adapt it continuously so that the whole organism would prosper.

NEXT GENERATION E-MANUFACTURING SOLUTIONS

A good starting point to illustrate the possible future challenges of agent-based manufacturing is from user requirements or from current impressions of experts in the field of future manufacturing information systems, often referred to more fashionably as next generation e-manufacturing solutions. A recent white paper by Interwave Technology [1] aims to synthesize the "myths, morphs, and trends" in this field, showing how the need for an effective MES layer, which might seem to have lost importance in the e-commerce age, is still one of the main priorities in manufacturing. Specifically, the following myths often quoted by manufacturing companies are introduced and discussed:

- It is far from true that new ERPs can manage real-time data. Whatever ERP vendors say, an ERP system is not designed to handle the real-time, process data-intensive world that characterizes an execution system, and an MES layer is still required.
- It is not true that an MES is an "add-on" layer of a shop floor system. If production is performed, MES functionalities (automatized or not) must already be present.
- It is true that new technologies are the key to successful MES.
- It is true that MES software is now an "a la carte solution," shifting the balance toward product configuration rather than customization.
- It is far from true that an MES will not impact the business processes of a manufacturing company. The proof is related to the fact that concepts like lean manufacturing and build-to-order models can only be implemented when business processes and plant-side systems are synchronized and optimized.
- One of the major mistakes that can be made is to believe that bypassing the so-called overhead tasks, such as project management, detailed operational design, documentation, and training, can save money because, after all, they have proven to be the keystones of the most successful projects in manufacturing companies thus far.

The same white paper reports three ways that MES is "morphing" to better serve today's rapidly changing manufacturing environments:

- MES is going mainstream, that is, it is increasingly identified as a key required component of the corporate supply chain. Some proof is found in the following:
  - Industry models, such as the Supply Chain Council's SCOR; MESA; ISA-95's; and AMR's collaborative manufacturing execution, incorporate MES as one of the key strategic and integral parts of a B2B/B2C enterprise strategy, and they identify MES as the enterprise data engine that will permit supply chains truly to run in real-time.
  - A common language is emerging for vertical and horizontal integration of the supply chain in the form of new standards (SCOR, MESA, S88, ISA-95, etc.), embracing XML to address aggregation of process, product, and plant data. These standards are earning broad consensus in the user community; this "morphing" will simplify applications and interface design, maintenance, and management of change.
  - There is now an unprecedented ease of information access for business decision support. Intranet/Internet technologies and the integration of related LAN, WAN at the different levels of the business process have provided enhanced access to information inside/outside the company's boundaries, as well as a way to integrate software functions written in any language and running in any operating system/hardware component. In addition, wireless technology in plant LANs and joint use of thin client terminals will allow delivering real-time information access when and where needed.
- MES is now an essential thread in the fabric of manufacturing. Some key MES functionalities, such as optimizing the design/engineering process and trending performance, are now main components of a shop floor system because it is reckoned that they can sensibly improve the overall manufacturing agility.

As a conclusion of the white paper by Interwave Technology [1], the following trends are identified:

- MES will become the engine behind inter/intramanufacturing enterprise collaboration and high-velocity supply chain performance.
- Multisite MES deployments (such as the one needed by PS-Bikes) will become increasingly necessary.
- Due to specific regulations, MES will become standard in medical device and biotech industries.
- Because of genetically modified foods, MES will become standard in the food and beverage industries.

- Manufacturing information visibility will be enhanced due to new technologies such as MS.NET and progressively integrated Windows/Web Forms, as well as on a comprehensive set of different hardware, including PDAs and cellular phones.
- The cost to develop, maintain, and integrate MES solutions will drop by 30% over the next 5 years as a result of the application of standards to prepackaged domain-based software.
- XML Web-based systems will be an essential feature of MES products for the years 2004 and 2005.
- Business change management will be enabled by specific "intelligent" applications.
- MES will raise the average acceptable plant efficiency benchmark from 60 to 70% to over 90% by 2006.
- MES will prevent the export of the U.S. manufacturing base.

Some reflection is needed at this point. Although the role of MES may be overemphasized in [1], and some points may be reasonable but not demonstrable, it is quite credible that MES will nonetheless play a fundamental role in next generation e-manufacturing. Throughout this book, the authors have reiterated their belief that one of the key roles of agent-based manufacturing to add agility to an enterprise (requesting it) lies in MESs. In fact, an efficient MES is still a request, and nearly all the myths, morphs, and trends quoted previously do not argue against, but rather, generally support, agent-based MES. Agent-based manufacturing ought to seize the opportunity to satisfy this request. The following sections briefly discuss, also in relation to the MES myths, morphs, and trends cited earlier, some important aspects upon which research on agent-based manufacturing should focus in the near- and in the medium/long-term future.

WHAT MUST BE REFINED IN AGENT-BASED MANUFACTURING

In this section, some key indications as to what must be refined over the next few years are presented. In keeping with the authors' purpose, this section aims to show which improvements should be expected in the near future and which areas need urgent work.

As a matter of fact, although agent-based systems are becoming increasingly well understood, MASs are not [2], and MAS development and application in manufacturing (or lack thereof) is a typical example of this fact. Wooldridge and Jennings [2] identified several categories of common problems specifically related to agents as a software engineering paradigm. These problems, which are nearly still present, reflect the main pitfalls facing the agent system developer; although they are often stressed from a software engineering perspective, they make a good starting point for the assessment of the errors in agent-based manufacturing implementation and, as a consequence, what must be urgently refined in agent-based manufacturing approaches.

Specifically, software engineers are likely to be subject to two major "political" pitfalls in a corporate environment: they may tend to "oversell" agents and often risk becoming dogmatic about them. Both of these attitudes are related to the need for an effective justification to use agents. This is more than true in manufacturing, especially when the many existing reliable alternatives from technological and methodological standpoints are taken into account. What urgently needs to be refined in agent-based manufacturing are the timing and setting when and where agents and MASs should be applied. In the authors' opinion, agents and MASs have by now shown themselves to be readily applicable whenever agility is requested in the enterprise and whenever their introduction could be performed without jeopardizing the performance of the existing legacy system. In this respect, the ideas that have been reported throughout this book aim to reflect a step toward the definition of possible environments in which agents and MASs may live in a manufacturing information system, and MES is one of the major opportunities available.

Two other important pitfalls identified by Wooldridge and Jennings [2] are related to "management." These pitfalls arise from the fact that managers proposing an agent-based project seldom know why they want to use agents; too often, the choice is made because it is a nice-sounding word for marketing. In addition, managers are often seeking generic solutions, while agents and MASs are generally more suited for certain types of tailored applications. In manufacturing, the constraint is even stronger because it is quite evident that product configuration is often preferred to product customization (see, for example, the fourth myth reported later in the chapter). In addition, it seems that MESs will become specialized for certain industrial categories (medical, biotech, food and beverage; see the trends in the second section). What appears to be needed urgently is an effort to produce practical exemplifications in specific manufacturing fields, for instance, in some MES requested for primary or support functionalities. In this context, agent-based simulation can help to analyze scenarios and to select an appropriate MAS architecture for an effective enterprise solution. Even though it is a didactic and introductory approach, the PS-Bikes exemplification strives to fulfill this role throughout this book.

Developers, on the other hand, may be beset by "conceptual pitfalls." In this regard, Wooldridge and Jennings [2] cite the belief in agents as a technique that will provide an order-of-magnitude improvement in software development, and the oversight that agents are just software — specifically, multithread software. These hazards hold in manufacturing as well, and specifically for multithread aspects; problems such as synchronization, mutual exclusion for shared resources, deadlock, etc. are felt all the more due to the soft/hard real-time requirements that some applications may have. On the other hand, extending the concept of agent-based manufacturing also to holonic manufacturing, agents should be viewed as something more than software: they should be seen also as a paradigm that allows a new modeling approach to planning, scheduling, and control processes.

Thus, what must be refined in an agent-based manufacturing approach is the methodological *a priori* cost/benefit evaluation of whether it is really the case to adopt this solution instead of traditional reliable techniques; if this assessment supports the use of agents, further methodologies ought to support the start of an accurate design phase followed by simulations to study the behaviors of the system with respect to agility performance and to classical multithread problems. As shown in Chapter 2 through Chapter 4, despite the wealth of literature on the subject, no methodologies or techniques seem to have reached a mature enough stage to be defined as the correct pathway toward agent-based manufacturing systems.

Wooldridge and Jennings have also singled out "analysis and design" drawbacks [2] in agent-based software. Specifically, they recommend using conventional software technologies; adopting standards; exploiting concurrency in design; taking into account legacy systems; and so on. Such recommendations are evident in manufacturing applications more than elsewhere and, again, would need to be supported by a standardized sequence of design/simulation/implementation steps.

At the agent level, the same authors [2] identified the three following pitfalls: the wish to implement yet another original agent architecture in every new implementation; the use of excessive artificial intelligence within an agent; and, in some cases, the use of no artificial intelligence at all. This also holds true for the several agent architectures proposed for manufacturing applications by the literature, even more so if the concept of holonic manufacturing is taken jointly into account with agent-based manufacturing, as it is throughout this book. Again, a standard seems to be the refinement that can shrink this problem. As regards the presence of artificial intelligence, the problem is perhaps different in the agent-based manufacturing application because it is a matter not only of artificial intelligence but also of the interaction of many other disciplines such as operations research, control systems, game theory, etc.

Specifically, this book has throughout identified two categories of agents in manufacturing applications: *business component agents* and *synthetic social agents*. The main feature of the first type is to fit exactly into one task or a subset of tasks requested to carry out a business process, whereas the second type is usually introduced to handle complex decisions, decentralizing the decisional capabilities among the different actors of the decision process and modeling this process by means of a social collaboration/competition paradigm. The next few years will likely see the progressive separation of these two categories in manufacturing implementations.

At an MAS level, Wooldridge and Jennings [2] identified five potential pitfalls: using too many agents; using too few agents; using yet another MAS original infrastructure for each implementation; allowing too much freedom in the interaction among agents; and underestimating the use of a proper MAS design structure. Again, these problems are admittedly also present in MAS manufacturing applications; however, although Wooldridge and Jennings' work [2] is nearly 5 years old, a solution to them will unlikely be found in the next few years.

Users can give still another, perhaps more imposing, opinion of what must be refined in agent-based manufacturing. Thus, to make this section more incisive, the key indications of urgent requirements in agent-based manufacturing cited earlier, together with additional examples, are expressed hereinafter as aspirations and prerequisites underpinning PS-Bikes' decision to adopt agent-based solutions.

In fact, PS-Bikes' managers are quite eager to "agentify" their company, and the technical background, as well as the examples of possible applications given in the previous chapters, also with reference to their enterprise, have nearly convinced them that MAS technology can guarantee agility in production needed for peak performance in today's society. However, due to the lack of a sufficient number of testbed applications, PS-Bikes' managers wish to receive additional feedback on the aspects outlined in the following subsections, which, in their opinion, need further refinement.

Standards

PS-Bikes is burdened by the need to comply with a myriad of standards, as are nearly all manufacturing enterprises. Research on agents does not seem overly concerned about the idea of providing a standard. Is this really true? What are the emerging standards for agent-based manufacturing?

"Standards" is an obsessive term that has been repeated relentlessly throughout this chapter and the whole book. Currently, the only standard that can be referred to is IEC 61499, mentioned in Chapter 3. However, as mentioned in Chapter 6, although the IEC 61499 standard can be viewed as the first step for holonic manufacturing, the same standard needs to be "dressed up" by an agent/MAS architecture. FIPA, cited even more throughout this book, will likely become the candidate framework to answer to this challenge: some interesting work, for example, Ulieru [3], seems to be proceeding in this direction, as do recent FIPA-compliant applications developed using Jade middleware.

In addition, agents and MASs are likely to be immersed in their "dialogues" with XML schemas for manufacturing, with emerging communication standards for pervasive and wireless technologies. Moreover, so that agents succeed as MES candidates in manufacturing information systems, agreement should be reached with ISA-95 and/or MESA-11 standards, as well as with the ISO/IEC 15408 standard [4] for information system security.

As regards agent technology in general, the AgentLink project predicts that in the near term (2003 to 2005) [5], communication languages based on standards such as FIPA will be increasingly used. However, interaction protocols will remain nonstandard and MAS development will increasingly use top-down methodologies, such as GAIA, or middle-out methodologies supporting applications based on service-oriented architectures.

In conclusion, although it is reasonable that PS-Bikes' agent-based manufacturing strategy ought to follow this trend, no consensus standard can be guaranteed at the moment.

Integration with Current Methodologies/Technologies: the Need of a la Carte Solutions

> There seems to be a wide gap between agent-based and traditional manufacturing. Is it a gamble to support the idea of a completely agent-based manufacturing information system, or will a specialized MAS be designed to be fully integrated with existing legacy systems? For example, in PS-Bikes, certain functions are not present in the information system yet; CRM in a new e-commerce functionality and workflow management might be enough for a preliminary experimental agent-based implementation. Can PS-Bikes rely on any a la carte agent-based solutions that are ready to be configured, or must they venture into customized options that demand deeper investigation?

Throughout this book, the authors have defined the notion of "agentification" of an enterprise as an ongoing process in which agents aim to enhance an existing legacy system. The few existing industrial applications seem to corroborate this view. *Ab initio* approaches are likely to be adopted when a future MAS is able to be truly proactive, evolving autonomously within the enterprise with a proper auction-based dynamic protocol.

If a sort of MAS4MES is to be defined in the next few years, specific applications should be taken into account for specific manufacturing sectors, also following the trends cited in the section on next generation e-manufacturing solutions. According to Luck et al. [5], in the near future, commercial demand for closed multiagent systems will likely be substantial because of the security concerns that arise from open systems. In the mid-term (2006 to 2008), open systems will typically be specific for particular application domains, such as e-commerce or bioinformatics, while bridge agents, able to translate between separate domains, will also be developed.

At the moment, as far as the authors know, no a la carte configurable agent-based solution is ready for any specific domain (let alone bike manufacturing).

Bidding vs. Executing: the Role of Simulation

> There is a lot of work to do in PS-Bikes. Are you sure that your agents work, without spending their time "to chat and to bid?" For example, regarding scheduling problems, are you sure that agent bidding-based scheduling is robust and always converges without any deadlocks, loops, and any way quickly? Is it possible to quantify — for example, through the simulation of some reference scenarios — the agility that PS-Bikes can earn by introducing agent-based technology?

The successful application of agent-based manufacturing is necessarily preceded by a cost/benefit analysis and, when an MAS approach is convenient, followed by a specific design/simulation implementation exercise. The performance of current auction protocols present in agent manufacturing should always be tested before implementation, and worst-case scenarios should always be simulated. However, MAS manufacturing is still lacking any sort of quality-control procedure.

Simulations, which have been investigated in Chapter 4, may be an advantage by themselves. For example, the AgentLink project [5] reports some successful agent-based simulations: ant-inspired agent-based simulations of complex supply chains have been used by EuroBios to assist logistics analysts and plant schedulers at Air Liquide in making better decisions; at Southwest Airlines, agent-based simulations of cargo routing revealed many missed opportunities to load cargo and enabled a 75% cut in the multiple handling of freight and an increase of U.S. $10 million in revenue.

Costs and Time

> How much does an agent and/or an MAS cost? How long does it take to install an MAS?

These quite simple questions are perhaps the most difficult to be answered. The easiest, most often heard reply is that these questions are nonsense because their answers depend on the application, the technology, and so on. Admittedly, an effort to reply to these questions should be made in order to raise the credibility of agent-based manufacturing as a candidate for real applications; it is difficult, however, to quantify these aspects at present, due to the scarcity of testbed applications. Nevertheless, some indirect estimates of price ranges that make agent-based manufacturing competitive as an MES application should be given.

A recent review on MES [6] reports costs of real applications range from $70 to $815 k, with $500k a rough average. Electronics, chemical, metallurgic, and pharmaceutical industries have implemented the heaviest MES projects thus far. In general, as seen in the trends of next generation e-manufacturing solutions, the cost to develop, maintain, and integrate MES solutions ought to drop by 30% over the next 5 years.

An MES project (including selection, request, analysis of requests, contract, design, implementation, test, etc.) can last from 3 months to 2 years, although implementation on its own should take from just 1 day to 1 month. However, it should be borne in mind that choosing an MES system is likened to marriage, in which divorce can prove painful. If MASs wish to flourish in MES, costs and timeframes should fall into these ranges in order to be competitive.

From another perspective, the feeling that agents are an outright business is evident from reports forecasting that "agents will generate U.S. $2.6 billion in revenue by the year 2000 [7]." The authors of this book are unable to substantiate this claim, but the growing attendance at meeting and workshops on the topic clearly bears out companies' attraction to agent-based manufacturing.

WHAT HAS YET TO BE DONE IN AGENT-BASED MANUFACTURING

Despite the urgency of the required refinements described in the previous section, some of them will probably not be achieved even in the next few years. Broadening the horizon of research activities into the next decade, this section aims to introduce the most promising fields of research entering on agent-based manufacturing.

In a recent survey more specific for manufacturing scheduling, Shen [8] foresaw the following priorities, opportunities, and challenges:

- Negotiation mechanisms and protocols
- Integration of planning, scheduling, and control
- Integration of agent-based and traditional approaches
- Combination of individual solving and coordination–negotiation
- Setting of benchmarks
- Theoretical investigation of methodology

Due to the central role of scheduling problems in manufacturing, the discussion presented next follows this taxonomy, extending the argument to planning and control when the case arises.

Negotiation Mechanisms and Protocols

According to Shen [8], the increasing use of bidding-based negotiation protocols requires research and development of more sophisticated negotiation mechanisms and protocols; specifically, combinatorial market-based negotiation protocols seem to be of much interest for the near future.

In fact, a lot of work still must be done as far as the design of interactions among agents is concerned; however, an interdisciplinary approach reproducing methodologies studied in different contexts and sciences should be used in order to avoid starting from scratch and reinventing the wheel. In 1997, Kraus [9] already felt this need and proposed practical examples in which negotiation and cooperation should not be taken into account as a problem limited to distributed artificial intelligence*; operational research; control systems; information systems; or any other discipline. In his tutorial, he proposed examples related to different disciplines that can be used to interpret agent interactions in an MAS.

The first set of examples proposed was related to the application of game theory techniques to multiagent environments. This kind of approach is recommended when a small number (less than a dozen) of agents are self-motivated and try to maximize their own benefit. The active entity is

---

* As introduced in Chapter 1, research in distributed artificial intelligence is divided into two basic classes: distributed problem solving (DPS) and multiagent systems (MASs). In general, cooperative agents belong to the DPS class, while self-motivated agents belong to the MAS class. However, due to the interdisciplinary approach of this book and to the fact that the distinction between these two approaches is not always so clear-cut in more complex problems such as organizational and production systems, the term "MAS" has been used throughout this book to define any system with two or more agents, independently of whether they cooperate or not.

a player, which in game theory is modeled according to one of two main types: noncooperative models or cooperative, where actions are respectively individual or joint among groups of players. The more than 20-year-old alternative offers model [10] is cited as an example of a strategic bargain model designed to assess what the gap is and what has yet to be done in the application of game theory to MAS design. The five aspects that follow already seemed relevant in 1997 [9], but they still deserve to be summarized:

- Choice of a strategic bargain model applicable in MAS design. One important aspect appears to be the use of models, as in Rubinstein [10], which take into account the passage of time during negotiation. The time it takes to reach an agreement is a very important factor for at least two reasons: the cost of communication and computation time spent on the negotiation, and the loss of unused information due to the block of information flows awaiting the decision. Recent work [11] seems to have taken up on this trend.
- Matching MAS scenarios with the game-theoretic definitions of the chosen model. This means identifying the players and covering them with specific agents, as well as the agreements that must be reached.
- Identification of equilibrium strategies by formalizing assumptions appropriate for MASs and for the related applications. For example, all agents sustain a loss over time; there is a large but finite set of agreements; there are some agreements that are better for all agents rather than opting out of the negotiations.
- Development of low-complexity techniques for researching appropriate strategies, mainly related to MAS situations in which the MAS designer cannot provide the agent with a negotiation strategy in advance. To construct strategies that can be proven to reach equilibrium can be done only when the set of possible agreements can be defined. There is a need to develop low-complexity techniques for these situations, and this element has been faced rarely by game-theory literature.
- Provision of utility functions. In game theory, the players' utility functions or preferences are important features, and each player knows its utility function (and has some knowledge of the utility function of its opponents). This is not so obvious in an MAS.

The second set of examples [9] was related to the application of classical mechanics to large-scale agent systems. In very large (hundreds) agent communities, such as the Internet, negotiation methods are typically computationally too complex and time consuming, and require a heavy exchange of messages. Physical models of particle dynamics have proved useful in such settings, using a mathematical formulation to describe or to predict the properties and evolution of different states of matter (an example applied to goods transportation is reported in Krauss [9]).

The third set of examples was related to the application of operations research techniques [9]. For example, task allocation among agents may be approached as a problem of assigning groups of agents to tasks; therefore, the partition of the agents into subgroups becomes the main issue and the problem can be viewed as a set partitioning problem (SPP) [12]. The recommended procedure to approach MAS design from an operations research vantage point is to recognize the problem as a traditional operations research problem, to find a related distributed formulation, and to adapt it to an MAS environment, also through the introduction of utility functions used by agents.

Finally, the fourth set of examples was related to the application of informal models of behavioral and social sciences to automated agents, which are useful specifically when there is the need to interact with humans [9].

The survey by Krauss therefore seems all the more up to date, also taking into account (as assessed by Luck et al. [5]) that no auction process will be standardized in the near future. In conclusion, designers of agent-based solutions will need increasingly to exploit, as a first step, traditional techniques coming from other fields in an interdisciplinary approach. In a subsequent step, it is likely that these approaches will become progressively specialized for agents and MASs, exploiting characteristics such as dynamic adaptation of the auction method — a sort of genetic, societal evolution of the auction method and the number of the agents involved in the bargain.

Integrating Planning, Scheduling, and Control

Shen [8] observed that agent-based approaches provide a natural way to integrate manufacturing process planning, scheduling, and execution control; specifically they can provide the possibility of simultaneous optimization of process planning and manufacturing scheduling. However, this integration aspect requires much more research on its formal modeling.

In fact, the need for agility — horizontal integration of the enterprise within the supply chain and as vertical integration between management and the shop floor — has been stated several times throughout this book, implying MES as a natural application for MASs. According to a recent presentation by Prabhakar and McClellan [13], the evolution of integration in collaborative manufacturing increases the complexity of enterprise/business processes following the steps sketched in Figure 7.1. This evolution has started from the first need to move data inside and outside the

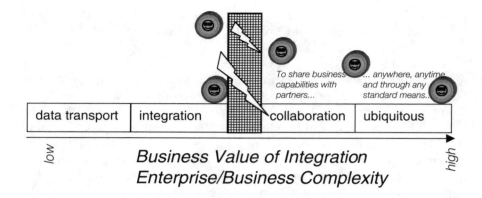

| data transport | integration | collaboration | ubiquitous |

low → *high*

Business Value of Integration
Enterprise/Business Complexity

Figure 7.1 Evolution of the Need of Integration in Manufacturing Enterprises

enterprise, to the need of data, application, and process integration. The current needs and requests of collaborative manufacturing are for "collaboration," to share business capabilities with partners, and for "ubiquitous processes" in order to share these capabilities anywhere, anytime, and through any standard means.

Although of proper technology to satisfy these new emerging integration needs is available, we are experiencing nowadays an "impasse" that, like a wall, stops the integration evolution of collaborative manufacturing. This impasse is probably due to the anxiety that this evolution could represent the introduction of an unsustainable degree of complexity of enterprise/business processes, but also to the lack of proper models and methods to introduce them. An MAS can represent the model and the method to break and to go beyond this wall.

The adoption of virtual manufacturing architectures needed for the survival of European SMEs, as well as their continuous need for updated information (see, for example, the scenario proposed by Luck et al. in section [5]), should also be related to the probable collapsing of the traditional three-layered architecture into one monolithic level. "Monolithic" should be related only to the unification of functionalities that, contrary to what its name suggests, would reflect a highly reconfigurable, dynamic, distributed architecture, i.e., an agile architecture. This has been shown throughout this book to be ready to be effectively implemented as an agent-/holonic-based manufacturing system. However, for many sectors in which agility is not required, traditional approaches will continue to suffice.

Integrating Agent-Based and Traditional Approaches

According to Shen [8], because bidding-based approaches emphasize flexibility and responsiveness rather than the optimality of solutions, they are more suitable for on-line* rescheduling. On the other hand, approaches to search such as genetic algorithms and simulated annealing, which focus more on the optimality of solutions, are more suitable for advance scheduling. In this respect, several possibilities of research are indicated by Shen [8]: for example, shop floors that require advance and on-line scheduling could combine some of these approaches; integrating agent-based approaches with other traditional approaches (for example, fuzzy logic, artificial neural networks, Petri net-based coordination, etc.) may be another interesting research aspect.

In fact, in an agile enterprise, advance and on-line rescheduling will be increasingly demanded, and it is likely that the contract protocol within an enterprise or among a cluster of enterprises forming a virtual enterprise will evolve according to a sort of evolutionary self-learning algorithm (which in some way could exploit ideas from the field of natural optimization algorithms like genetic algorithms and simulated annealing).

In the long-term outlook (2009 onwards), AgentLink [5] predicts the development of open multiagent systems spanning multiple application domains that involve heterogeneous participants developed by diverse design teams. According to AgentLink [5], agents seeking to participate in these systems will be able to learn the appropriate behavior for participation while doing so, rather than needing to prove adherence before their admission. As regards their integration with traditional methodologies, similar studies are already present in the literature; see, for example, Lin and Norrie [14] on Petri nets, and Mulieru and Norrie [15] and Maione and Naso [16] on fuzzy set theory.

Combining Individual Solving and Coordination–Negotiation

The presence of an obvious trade-off between solving at the individual agent level and the coordination–negotiation scheme at the system level has been pointed out by Shen [8]: a promising and challenging research aspect is how to combine them using integration.

In fact, these problems have hampered agent and MAS frameworks since their advent. For example, Castelfranchi and Conte [17] examined some crucial limits of game theory approaches often used in MASs: the absence of a specific notion of cooperation, giving more emphasis to goals rather than to behavior, and the absence of a general model of social influence. In addition, designing effective forms of cooperation among local autonomous agents, without contradicting in some way the

---

* In this book, we have chosen the term "on-line scheduling" rather than "dynamic scheduling" as used by Shen, leaving dynamic scheduling to other significance. See Chapter 3 for major details.

true MAS design principles based on the absence of any hierarchical form of supervision, is against the most intuitive MAS model because it seems to limit autonomy of agents with the introduction of coordination rules. On the other hand, giving the responsibility of coordination to one or more agents, which are also the repository of the current state of the utility functions of the other contracting agents, can give more stability to the auction algorithm; as a result, this kind of approach will continue to be applied in manufacturing for many years to come. However, the long-term future is likely to produce new research solutions in which agents will autonomously find their proper role in the agent society, with the capability of influencing the utility functions of the other agents in a network of influences that will evolve in time and complexity.

Benchmarks

Shen has [8] remarked that benchmarks are needed to compare different agent-based systems as well as to compare these systems with others using traditional approaches. In fact, the setting of benchmarks has also been urged in the most recent works. For example, Maione and Naso [16] compared the performance of approaches on a detailed simulation model of a hypothetical manufacturing system recently proposed as a benchmark for an MAS [18]. Cavalieri et al. were among the first to refer to the need for a common benchmark [19] because many authors often fail to provide sufficient detail on their design hypotheses and on the structural charac-teristics of the manufacturing system. Again, no standard methodology based on a common reference point is currently available, and pains must be taken to define benchmarks with a sufficient degree of complexity.

Theoretical Investigation of Methodology

Finally, Shen [8] has indicated an urgent need in the field of theoretical investigation of methodology, including implementation methodology, to consolidate current research results and facilitate implementation of real industrial applications. In fact, theoretical investigation of methodology covers and will always cover many facets, some of which have been touched on in this book. Among the others, references modeling frame-works deserve mention. Existing modeling frameworks for manufacturing system control can be classified as hierarchical, heterarchical, or hybrid control (see Chapter 3). Hybrid control frameworks seem to be the best candidate for next generation MASs, and the work by Heragu et al. [20] can be considered an important theoretical investigation of methodology per se and as a model of scientific work to be followed by other researchers in this field.

According to the work of Jennings and Bussmann in which they discuss the suitability of agent-based control systems for engineering complex systems [21], two key pragmatic issues will determine whether agent-oriented approaches catch on as a software engineering paradigm: the degree to which agents represent a radical departure from current software engineering thinking and the degree to which existing software can be integrated with agents. It is thus clear that theoretical investigation of methodology will also deal with approaches to make the "agentification" of an enterprise a smoother, although growing and evolutionary, transition process, facilitating in this way the implementation of real industrial applications.

PS-BIKES AND ITS CHALLENGING "AGENTIFICATION" PROCESS TO ACHIEVE AGILITY

The future challenge for PS-Bikes is to "agentify" its manufacturing practices through a smooth, growing, and evolutionary process. To achieve this, PS-Bikes is defining the following technical tender (reported in its draft form) to request the following activities:

- Design and implementation of an MAS platform to enhance workflow management and customer relationship management (CRM), supporting the development of an MAS to support its imminent e-commerce activities. The MAKE-IT approach described in Chapter 2 seems to be reasonable for workflow management, as does the design/implementation approach described in Chapter 5 following the PASSI methodology. The JADE middleware seems to be reasonable for CRM, but the question of standards, which may also have an important commercial impact, should be taken into account more specifically as regards the use of XML standard schema for manufacturing; and the compatibility with ISO/IEC 15408 standard as regards security and of the ISA-95 as regards interaction with the shop floor.
- Design and implementation of an MAS-/holonic-based system in the new northern plant. The system will conform to the ISO/IEC 15408 standard function blocks and design. The scheduling problems will be solved on-line by suitable contracting/coordination algorithms distributed in a hybrid framework.
- Design and implementation of an MAS for virtual manufacturing MES. PS-Bikes wishes to substitute the current functionalities of its MES with an agent-based MES enabling management of the two plants as if they were one virtual facility. In addition, this functionality should be extended to other SMEs with which PS-Bikes wishes to cooperate

year.month

Activities 1st year	1.1	1.2	1.3	1.4	1.5	1.6	1.7	1.8	1.9	1.10	1.11	1.12
WF MAS	cost/benefit analysis			design			simulation			implementation		
CRM MAS	cost/benefit analysis			design			simulation			implementation		
HOLONIC MAS	cost/benefit analysis						design			simulation		
VIRTUAL MAS4MES	cost/benefit analysis						design v.1			simulation v.1		

Activities 2nd year	2.1	2.2	2.3	2.4	2.5	2.6	2.7	2.8	2.9	2.10	2.11	2.12
WF MAS	simulation/implementation refinements											
CRM MAS	simulation/implementation refinements											
HOLONIC MAS	implementation			simulation/implementation refinements								
VIRTUAL MAS4MES	implementation v.1			design v.2			simulation v.2			implementation v.2		

Figure 7.2 Simplified GANTT Chart of the Agentification of PS-Bikes

in the near future. Again, compliance with standards should be taken into account in the development. The transition toward the new MES should be achieved gradually; in a first stage, the use of a new MAKE-IT agent with a sales supervisor role, which handles the tasks of monitoring the arrival of new orders from the Web and announces them to the sales agencies of the two plants, should be implemented.

- Prerequisites for each implementation and methodological approach. Before starting each of the implementations cited earlier, the project should be justified with an adequate cost/benefit methodology. Each MAS design should follow an adequate methodology, with preference for the GAIA or the PASSI model. A simulation phase should follow, preferably adopting assessed testbed benchmarks or scenarios. The implementation phase will follow, preferably adopting available open source and/or commercial technologies; the use of a standard of reference (e.g., FIPA), and/or the use of available middleware (e.g., Jade) should be preferable. Testing will follow, and an iterative procedure to enhance the MAS should be provided, in which the simulation should be reliable enough to reproduce the current system behavior.
- The sequence of activities achieving the agentification of the PS-Bikes is illustrated in Figure 7.2 as a simplified Gantt chart. Four project activities are foreseen in relation to the preceding points: WORKFLOW MAS (WFMAS); customer relationship management MAS (CRM MAS); holonic MAS; and virtual MAS for MES (VIRTUAL MAS4MES).

CONCLUSIONS

Four overriding needs seem to fuel future investigations and developments in order to exploit agent-based systems systematically as a reliable solution in manufacturing:

- A methodology for analyzing the MS (e.g., MES) requirement that drives the definition of the MAS specifications
- A computer-aided software engineering (CASE)/simulation testing environment oriented to manufacturing applications that joins agent-based simulation and discrete event simulation facilities
- The availability of standard software components (wrappers) through which agents can be easily interfaced to physical devices (e.g., machines) and information system modules (e.g., DBMS) that are largely diffused in manufacturing systems
- The availability of *best practice* agent-based solutions to be considered as the reference starting point for the rapid customization of manufacturing applications such a la carte solutions offered as add-in modules of conventional ERP/MES packages (at least for SMEs) to enhance their response to the "agility" issue

As a very last consideration, it is important to delineate the framework in which planning, scheduling, and control problems must currently be faced in modern manufacturing systems because it dictates to a great extent the suitability of an agent-based system approach. More specifically, the reader's attention should be focused on the following questions:

1. What is the manufacturing system of reference?
2. Are single or multiple entities involved in the planning, scheduling, and control activities?
3. Are the decisions made in some centralized way, or can their responsibility be distributed?
4. Which decisions can be made in advance (off-line) and which in real time (on-line)?
5. What algorithms or methods can be used to make optimal decisions that fit with the information actually available?

Let us briefly try to find the answers to these points:

1. A wide variety of manufacturing systems can be found. A manufacturing system may be an industry with a single plant producing a few chip mass products that do not need customization according to a make-to-stock philosophy, and the suppliers a few big raw material distributors (e.g., the production of one kind of food or housekeeping product). At the other extreme, a manufacturing system may be a network enterprise, in which high-quality products are manufactured as the result of cooperation of a set of specialized industries, with one or more plants each; the products are highly customized and produced on a make-to-order (assembly-to-order

or design-to-order) basis. More details can be added to separate the two production scenarios even more. However, the scenario sketched out is still useful to highlight several interesting points, some of which are not always trivial:

- The larger the number of inputs and outputs to a MS is, the harder the planning, scheduling, and control decisions become.
- An MS composed of many interconnected entities whose activities must be coordinated in order to reach at best the MS objectives is clearly more difficult to manage, for example, because first the plans and then the schedules must be coordinated.
- High-quality customized and more expensive products call for high-quality plans because errors in decisions can produce very negative economic effects.

The preceding considerations underline the need for a very effective planning and scheduling system that can rely on up-to-date information provided by an enterprise information system (IS). Information technology's current answer is highly interoperable ISs among the various actors in the production scenario and provided with analytical (decision) capabilities for all the aspects characterizing the manufacturing supply chain. Marketing keywords are supply chain management (SCM — more than "traditional" enterprise resource planning — ERP); customer relationship management (CRM), or Internet-enabled systems (B2B, e-procurement, e-commerce), which can be a module of an SCM system.

Why and when use an MAS? Company networks are not born as they finally become; they grow because the firm wants to expand its business or simply to survive. Large multinational companies are endowed with costly state-of-the-art ISs, which are extended progressively to include the SCM capabilities as needed. Such large industries usually operate with monolithic proprietary systems. Most of the manufacturing companies in Europe are significantly smaller, but they nevertheless usually have ISs. They can often represent an entity of an industrial framework composed of several companies used to operating with low-connectivity procedures.

This is a context in which MASs can best prove their worth. MASs provide the operational infrastructure to make separate entities exchange information and plans and, finally, to interoperate. MASs are not invasive, but can instead act as middleware with respect to the information or legacy systems already adopted in the companies. Thus, MASs can be considered a technology enabling the interconnection of companies in a partner network at lower cost that can evolve into a system more appropriately defined as a supply network — rather than a supply chain —

management system. The key lies in the scalability of an IS composed of distributed elements (the agents) able to interoperate efficiently. If an MAS approach is suited for planning, scheduling, and control within a single isolated MS, it is easily recognized that an MAS for planning, scheduling, and control in a network of companies may be no more than a matter of scale of the communication framework, on the condition that its architecture and the agents' roles have been appropriately designed.

On the other hand, it may be less evident that, in some cases, even the MS of the first kind introduced earlier, i.e., featuring a few simple products and limited interactions with suppliers and customers, can benefit from MASs. In these cases, MASs may not be devoted to planning or scheduling, which may be faced with consolidated, conventional procedures, but to the critical role of facilitating the procurement of the information on which the conventional planning and scheduling methods rely. MASs can interconnect the company with the supplier and distributor or retailer systems by substituting traditional communications based on phone or fax to render continuously available up-to-date information on customer demand or supplier offers.

2. Part of the answer to this question has been provided in the preceding answer. However, it is worth stressing that, in the presence of multiple entities, the activities of planning, scheduling, and control become more difficult (and at the same time critical). They can still be effectively supported by an MAS approach. To show this, it should be borne in mind that the responsibility for these activities at the three levels of an MS (namely, the planning, execution, and operation levels), in the ISA and MESA models cannot always be assigned to a single entity, but to a set of distributed entities operating at the same level. This may be the case of several production plants of an MS that share in part the product portfolio; planning and scheduling can be performed by *a priori* assigning each plant its production objective. Still better performance can clearly be achieved by considering the overall production capability of all the plants in an integrated way and defining a comprehensive plan (schedule) for the MS.

Often, this latter strategy, even if theoretically conducive to optimal performance, may be not practical for several reasons, e.g., computational complexity; lack of timely information; poor flexibility (robustness) in case of possible variations; and so on. The MAS approach to planning, scheduling, and control in this setting not only can underpin an intermediate strategy that improves the feasibility of the MS with respect to *a priori* fixing of separate decisions for the plants, but also can reduce the complexity of a wholly aggregate decision through a distribution policy. Note once again that this observation ignores details about the MAS architecture and the roles of the agents.

3. The distribution of decision capabilities and responsibilities is a well-known approach to tackling complexity. By the same token, often the only strategy to assure theoretically that the selected decision is optimal is to take all the aspects of a problem into account at the same time, i.e., to adopt a centralized approach. Sometimes, the responsibility for decisions is actually distributed among different entities because these latter correspond to different subsystems of an MS (or network enterprise) that do not have any hard influence on each other. As an example, one can consider a food industry with several plants, each devoted to a different food preparation (e.g., canned beans or frozen vegetables sold in paper cartons). The decisions about planning relevant to such plants are clearly decoupled because they respond to different demands and do not share production or transportation resources; as a consequence, scheduling decisions and control are also independent.

Distribution (decentralization) as a management and decision policy is meaningful at the least in the presence of subsystems with interconnections and influence relationships. For example, in planning and scheduling for multiplant production, the decision complexity can be reduced by considering smaller decision problems, each associated with a single plant. The point here is to find an acceptable (for MS executives) trade-off between an overall centralized production optimization and a simpler, more flexible and robust, but also approximate, distributed optimization. One must also bear in mind the two main aspects that further aggravate an overall centralized approach:

■ The lack of timely and reliable information about the problem (from demand information to that pertaining to the state of the production process)
■ The usually exponential increase of the time needed to find an optimal decision, which ensues from the increase in the number of variables that must be considered (the so-called problem dimension)

If an MS is able to handle these two tests in a satisfactory way, it is very likely that it needs neither a distributed approach nor an MAS (even if some issues about its responsiveness and flexibility could be raised). On the other hand, if at least one of these aspects is critical, then a distributed (in particular, MAS) approach can help. Note that an MAS is not the only distribution approach available;

however, once the appropriate MAS architecture has been chosen, it can support both aspects to the extent desired.

4. The previous point places in evidence the obvious (but sometimes neglected) importance of available information for correct decisions. The question here centers on the time at which such information is available. There are two main scenarios: the information is available either before the decision process can start or during its execution. For instance, if the data relevant to the customer orders that must be produced in the next production period (say, a week) are available in advance, the decision process that identifies the production orders (i.e., the production objectives for each single day of the week) can be performed off-line, before the beginning of the relevant production period. On the other hand, if only some of the customer orders are known in advance and the others arrive after the beginning of the production period during which they must be manufactured, then the problem is on-line because the decisions must be made in real time.

Clearly, the performance of planning and scheduling can be no better in on-line problems than in an ideally correspondent off-line problem for which all of the information is available in advance. Off-line problems are very often complex (i.e., computationally intractable), requiring very complicated and information-demanding, or fast and simple but far from optimal, methods. Although agents and MASs can be adopted to handle off-line problems — basically, as a decomposition (heuristic) approach — the benefits deriving from an MAS approach are more evident whenever an on-line problem must be faced. The reasons for this can be found in:

- The responsiveness of software agents to on-line (i.e., real-time) events
- The basic simplicity of on-line decision approaches that can be implemented more easily compared to a complex, even if distributed, off-line solution approach
- The ability of simple distributed agents to obtain updated information about the state of (part of) the system promptly

Agents can thus be called on to make decisions about the local management of resources depending on the occurrence of events.

Some readers may note the absence of the so-called dynamic decision problems among the preceding scenarios; a brief explanation follows to convey the authors' point of view about the difference between on-line and dynamic problems. Most of the time, the term *dynamic* is inaccurately used as an alias of on-line, when actually on-line means that at least part of the information needed for decision-making becomes available only during the decision process. Dynamic problems are characterized by some probabilistic information about the events and entities that, on-line, will affect the problem's solution. Such information allows, in general, the (off-line) statistical analysis of the performance of on-line policies (in some cases obtained in an analytical way, but most often through simulation). What must be underlined is that dynamic problems, as defined here, often simply turn out to be on-line ones because the probabilistic information on which the statistic analysis is based is imprecise and unreliable. Finally, from a practical standpoint, dynamic decision policies very often do not differ from on-line decision rules.

5. This last issue also allows us to summarize the concepts about the appropriateness of an MAS approach to planning, scheduling, and control problems in MS. If the context in which these kinds of decision activities must be performed allows the use of a satisfactory off-line, centralized, and optimal approach, change is not actually needed. Therefore, the answer to the last question is: if your optimization method is already thoroughly acceptable, you must wisely assess the suitability of an MAS approach. On the other hand, if you believe that the future of your system is not stable and that something (namely, more flexibility and agility) is needed to be ready to meet new market challenges, investment in an MAS deserves consideration.

REFERENCES

1. Next-generation e-manufacturing solutions: myths, morphs and trends, interwave technology, The White Paper Series, 2003, available at http://www.interwavetech.com/pdfs/ITI_WP_Myths_Morphs_Trends.pdf
2. Wooldridge, M.J. and Jennings, N.R., Software engineering with agents: pitfalls and pratfalls, *IEEE Internet Computing*, May–June, 20, 1999.
3. Ulieru, M., FIPA-enabled holonic enterprise, in *Proc. 8th IEEE Int. Conf. Emerging Technol. Factory Automation*, 2, 453, 2001.
4. ISO/IEC 15408-1, Information technology — security techniques — evaluation criteria for IT security, available at http://www.iso.org/
5. Luck, M., McBurney, P., and Preist, C., Agent technology: enabling next generation computing. A roadmap for agent based computing, 2003, available at http://www.agentlink.org/roadmap/index.html
6. Visser, R., MES vendor survey 2001–02, oral presentation, available at http://www.mesa.org, see also http://www.mescc.com/
7. Guilfoyle, C., *Intelligent Agents: the Next Revolution in Software*, Ovum Ltd., London, 1994.
8. Shen, W., Distributed manufacturing scheduling using intelligent agents, *IEEE Intelligent Syst.*, January/February, 88, 2002.

9. Krauss, S., Negotiation and cooperation in multi-agent environments, *Artif. Intelligence*, 94, 79, 1997.
10. Rubinstein, A., Perfect equilibrium in a bargaining model, *Econometrica*, 50, 97, 1982.
11. Fatima, S.S., Wooldridge, M., and Jennings N.R., An agenda-based framework for multi-issue negotiation, *Artif. Intelligence*, 152, 1, 2004.
12. Shehory, O. and Kraus, S., Task allocation via coalition formation among autonomous agents, in *Proc. IJCAI-95*, Montreal (Que), 655, 1995.
13. Prabhakar, R. and McClellan, M., Collaborative manufacturing, oral presentation, available at http://www.mesa.org/
14. Lin, F. and Norrie, D.H., Schema-based conversation modeling for agent-oriented manufacturing systems, *Computers Ind.*, 46, 259, 2001.
15. Mulieru, M. and Norrie, D.H., Fault recovery in distributed manufacturing systems by emergent holonic re-configuration: a fuzzy multi-agent modeling approach, *Inf. Sci.*, 127, 101, 2000.
16. Maione, G. and Naso, D., A soft computing approach for task contracting in multi-agent manufacturing control, *Computers Ind.*, 52, 199, 2003.
17. Castelfranchi, C. and Conte, R., Limits of economic and strategic rationality for agents and MA systems, *Robotics Autonomous Syst.*, 24, 127, 1998.
18. Cavalieri, S. et al., A benchmark framework for manufacturing control, in *Proc. 2nd Int. Workshop Intelligent Manuf. Syst.*, Leuven, 22, 1999.
19. Cavalieri, S., Garetti, M., Macchi, M., and Taisch, M., An experimental bench-marking of two multi-agent architectures for production scheduling and control, *Computers Ind.*, 43, 139, 2000.
20. Heragu, S.S., Graves, R.J., Kim, B.I., and St. Onge, A., Intelligent agent-based framework for manufacturing systems control, *IEEE Trans. Syst., Man, Cybernetics — Part A: Syst. Hum.*, 32, 560, 2002.
21. Jennings, N.R. and Bussmann, S., Agent-based control systems: why are they suited to engineering complex systems? *Control Syst. Mag., IEEE*, 23, 61, 2003.

INDEX